Springer Climate

Series Editor
John Dodson

More information about this series at http://www.springer.com/series/11741

Ross J. Salawitch • Timothy P. Canty
Austin P. Hope • Walter R. Tribett
Brian F. Bennett

Paris Climate Agreement: Beacon of Hope

Ross J. Salawitch
Department of Atmospheric and Oceanic
  Science
University of Maryland
College Park, MD, USA

Department of Chemistry and Biochemistry
University of Maryland
College Park, MD, USA

Earth System Science Interdisciplinary Center
University of Maryland
College Park, MD, USA

Austin P. Hope
Department of Atmospheric and Oceanic
  Science
University of Maryland
College Park, MD, USA

Timothy P. Canty
Department of Atmospheric and Oceanic
  Science
University of Maryland
College Park, MD, USA

Walter R. Tribett
Department of Atmospheric and Oceanic
  Science
University of Maryland
College Park, MD, USA

Brian F. Bennett
Department of Atmospheric and Oceanic
  Science
University of Maryland
College Park, MD, USA

ISSN 2352-0698   ISSN 2352-0701   (electronic)
Springer Climate
ISBN 978-3-319-83623-2   ISBN 978-3-319-46939-3   (eBook)
DOI 10.1007/978-3-319-46939-3

Printed on acid-free paper

This Springer imprint is published by Springer Nature
The registered company is Springer International Publishing AG
The registered company address is: Gewerbestrasse 11, 6330 Cham, Switzerland

To the children of the sixties ... the 2060s ...
may you be born when half of all energy
is supplied by sources that release
no atmospheric greenhouse gases
and may you live during prosperous times
in a world that has not experienced global
warming catastrophe.

Preface

On 11 November 2014, a remarkable event occurred. President Barack Obama of the United States and President Xi Jinping of China announced a bilateral agreement to reduce the emission of greenhouse gases (GHGs) that cause global warming by their respective nations. On 12 December 2015, a year and a month later, representatives of 195 countries attending the 21st Conference of the Parties of the United Nations Framework Convention on Climate Change meeting in Paris, France, announced the Paris Climate Agreement.

The goal of the Paris Climate Agreement is to limit the future emission of GHGs such that the rise in global mean surface temperature will be no more than 1.5 °C (target) or 2.0 °C (upper limit) above the pre-industrial level. The Paris Climate Agreement utilizes an approach for reducing the emissions of GHGs that is distinctly different than earlier efforts. The approach for Paris consists of a series of Intended Nationally Determined Contributions (INDCs), submitted by the world's nations, reflecting either a firm commitment (unconditional INDCs) or a plan contingent on financial and/or technological support (conditional INDCs).

The Obama–Xi announcement was instrumental in the framing of the Paris Climate Agreement. The INDCs submitted by the USA and China were built closely upon the November 2014 bilateral announcement. China and the USA rank number one and two, respectively, in terms of national emission of GHGs. Practically speaking, unified global action to combat global warming required these two nations to get on the same page.

Here we provide an analysis of the Paris Climate Agreement written for two audiences. The first audience is the bewildered public. Hardly a day goes by without some newsworthy item being reported on climate change. Often the stories are contradictory, tainted by parochialism, skepticism, and extremism by not only the conservative and liberal media but also the camps of so-called believers and deniers. Our book goes back to basics, outlining what is known and not known about climate change. If we have been successful, this book will enable readers to advance their own understanding of this topic, in a manner that will assist in the proverbial "separation of the wheat from the chaff" with regard to climate change.

Our second audience is the women and men who are charting the response of the world to the threat of global warming. As is clear from the title of this book, we believe the Paris Climate Agreement is truly a Beacon of Hope. The Agreement has been severely criticized by some scientists, even a few prominent in the field of climate change. In this book, we closely examine the behavior of the computer models commonly used to inform climate change policy. This examination will be eye opening to many. We urge policy makers to seek their own independent assessment of the veracity of the global warming projections that are being used to inform policy.

The heart of our evaluation of the Paris Climate Agreement is projections of global warming found using our own computer code, termed the Empirical Model of Global Climate (EM-GC). Calculations conducted in the EM-GC framework are the basis for our conclusion that the goal of the Paris Climate Agreement could actually be achieved, if the INDCs are fully implemented (conditional as well as unconditional) and if the reductions in the emission of GHGs needed to achieve the INDCs are propagated forward in time, with continuous decreases in the emission of GHGs until at least 2060.

This book emerged from a talk given by the lead author, at the January 2016 American Meteorology Society meeting. We thank the conveners of the meeting for giving our talk a prominent slot, which led to our work being noticed by Springer. We thank Zachary Romano of Springer Nature for his enthusiastic support throughout the duration of the project, as well as Susan Westendorf and Aroquiadasse JoyAgnes for their fantastic work during the production of this book. We appreciate the comments of many colleagues, way too plentiful to name, for constructive criticism of the emergent science from our EM-GC, as we gave talks at national meetings, small conferences, and department seminars.

This book emerged from a homework assignment, first given in September 2009, to a Numerical Methods in Atmospheric and Oceanic Science class at the University of Maryland. The assignment asked students to reproduce a figure involving multiple linear regression of global mean surface temperature from a paper written by Judith Lean and David Rind that had just appeared. Over the years, many students contributed to the development of our EM-GC code from its early root in this homework problem, which we sincerely appreciate. We especially thank Nora Mascioli, with whom we collaborated before she enrolled in graduate school at Columbia University. Three of us had the privilege of teaching a freshman Honors class on the Economics, Governance, and Science of climate change and two of us have taught a large freshman Introduction to Weather and Climate class. This book has benefited enormously from all we have learned from our students. We appreciate as well the collegial environment created by our colleagues, graduate students, and undergraduates at the University of Maryland.

For anyone who aspires to write a book, please know there will be a period of your life where "eat, sleep, and write" becomes the daily routine. The five authors sincerely appreciate our families and friends for their unwavering support during the long hours spent on campus. We greatly appreciate the time and effort of our eagle-eyed proofreaders, Heidi and Michael Hope and Gordon Dryden, who greatly improved the final manuscript by their fastidious attention to detail.

We appreciate the support of the NASA Climate Indicators and Data Products for Future National Climate Assessment (INCA) program, the sponsor of the research that led to this book. Our proposal was selected in response to the 2014 NASA Research Opportunities in Space and Earth Sciences INCA call. The material in this book reflects the views of the authors, and not those of NASA, the US Government, or our employer, the State of Maryland.

Finally, the figures used throughout the book are available electronically at http://parisbeaconofhope.org. This book is published under a Creative Commons License that permits use of figures, provided proper attribution is given. Annual updates will be provided for many figures on our webpage.

College Park, MD   Ross J. Salawitch
August 2016   Timothy P. Canty
                                                                       Austin P. Hope
                                                                   Walter R. Tribett
                                                                     Brian F. Bennett

Contents

Abbreviations and Acronyms

ΔRF	Difference in the RF of climate relative to a baseline, usually year 1750
ΔT	Difference between GMST and a baseline, usually years 1850–1900
γ	Climate sensitivity parameter, dimensionless
κ	Ocean heat uptake efficiency coefficient, W m^{-2}°C^{-1}
λ	Climate feedback parameter, W m^{-2}°C^{-1}
λ_P	Planck response function, 3.2 W m^{-2}°C^{-1}
τ	Lifetime of a molecule in the atmosphere
AAWR	Attributable Anthropogenic Warming Rate
AerRF$_{2011}$	Radiative Forcing Due to Tropospheric Aerosols, year 2011
AH	Attain and Hold
AH^{UNC}	Attain and Hold scenario, for which only unconditional INDCs are considered
AI	Attain and Improve
AI^{UNC}	Attain and Improve scenario, for which only unconditional INDCs are considered
AI$^{UNC+COND}$	Attain and Improve scenario, for which unconditional and conditional INDCs are considered
AMO	Atlantic Multidecadal Oscillation
AMOC	Atlantic Meridional Overturning Circulation
AMV	Atlantic Multidecadal Variability
Annex I	UNFCCC list of developed nations, prominently featured in Kyoto Protocol
Annex I*	UNFCCC list of developed nations, minus the US. This is our nomenclature, not that of UNFCCC
BAU	Business As Usual
BTU	British Thermal Units
CCS	Carbon Capture and Sequestration
CDIAC	Carbon Dioxide Information Analysis Center of the US DOE
CFCs	Chlorofluorocarbons: these gases deplete stratospheric ozone and warm climate

CMIP5   Climate Model Intercomparison Project Phase 5
CO_2-eq   Emissions of $CO_2^{FF} + CO_2^{LUC} + CH_4 + N_2O$, expressed in terms of CO_2-equivalent, found using GWPs for CH_4 and N_2O
CO_2^{EQ-IN} CO_2-equivalent emissions from individual nations
CO_2^{FF} Atmospheric CO_2 released by the combustion of fossil fuels, flaring, and the cement manufacturing, globally
CO_2^{FF-IN} Atmospheric CO_2 released by the combustion of fossil fuels, flaring, and the cement manufacturing, by individual nations
CO_2^{LUC} Atmospheric CO_2 released by land use change, globally
COP   Conference of the Parties
CRU   Climatic Research Unit of the Univ. of East Anglia, UK
DNB   Day Night Band
DOE   Department of Energy
ECS   Equilibrium Climate Sensitivity
EDGAR   Emissions Database for Global Atmospheric Research of the JRC, European Commission
EIA   Energy Information Agency
EM-GC   Empirical Model of Global Climate
ENSO   El Niño Southern Oscillation
EPC   Electricity Power Consumption
ESRL   Earth System Research Laboratory of US NOAA
GCF   Green Climate Fund
GCMs   General Circulation Models
GDP   Gross Domestic Product
GHG   Greenhouse Gas
GISS   Goddard Institute for Space Studies
GMST   Global Mean Surface Temperature
GWP   Global Warming Potential
Gt   Gigatonne, 10^9 metric ton
HCFCs   Hydrochlorofluorocarbons: these gases deplete stratospheric ozone, although not as strongly as CFCs; these gases also warm climate
HFCs   Hydrofluorocarbons: these gases do not deplete stratospheric ozone, but they warm climate
IEA   International Energy Administration
INDC   Intended Nationally Determined Contribution
IOD   Indian Ocean Dipole
IPCC   Intergovernmental Panel on Climate Change
JRC   Joint Research Center of the European Commission
LUC   Land Use Change
MLO   Mauna Loa Observatory
MLR   Multiple Linear Regression
MODIS   Moderate Resolution Imaging Spectroradiometer of US NASA
MW   Megawatt, 10^6 W
MWP   Medieval Warm Period
NASA   National Aeronautics and Space Administration

NCEI	National Centers for Environmental Information
NOAA	National Oceanographic and Atmospheric Administration
Non-Annex I	UNFCCC list of developing nations, prominently featured in Kyoto Protocol
Non-Annex I*	UNFCCC list of developing nations, minus China and India. This is our nomenclature, not that of UNFCCC
NPP	National Polar-orbiting Partnership of the US NASA, NOAA, and Department of Defense
ODS	Ozone Depleting Substance
OHC	Ocean Heat Content
OHE	Ocean Heat Export
ORNL	Oak Ridge National Laboratory of the US DOE
PBL	Planbureau voor de Leefomgeving; the Environmental Assessment Agency of the Netherlands
$pC^{EQ\text{-}IN}$	Per-capita release of $CO_2^{FF}+CO_2^{LUC}+CH_4+N_2O$, expressed in terms of CO_2-equivalent, by individual nations
$pC^{EQ\text{-}GL}$	Per-capita release of $CO_2^{FF}+CO_2^{LUC}+CH_4+N_2O$, expressed in terms of CO_2-equivalent, globally
pC^{GL}	Per-capita release of CO_2 due to the combustion of fossil fuels, flaring, and the cement manufacturing, globally
pC^{IN}	Per-capita release of CO_2 due to the combustion of fossil fuels, flaring, and the cement manufacturing, by individual nations
PDO	Pacific Decadal Oscillation
PFCs	Perfluorocarbons: these gases do not deplete stratospheric ozone, but they warm climate
PICR	Potsdam Institute for Climate Research
ppy	Per person, per year
RCP	Representative Concentration Pathway
RF	Radiative Forcing
SEDAC	Socioeconomic Data and Applications Center
SOD	Stratospheric Optical Depth
SPO	South Pole Observatory
TCRE	Transient Climate Response to cumulative CO_2 Emissions
Tg	Teragram, 10^{12} g
TSI	Total Solar Irradiance
UNFCCC	United Nations Framework Convention on Climate Change
VEI	Volcanic Explosivity Index
VIIRS	Visible Infrared Imaging Radiometer Suite, US NASA and NOAA

Contributors

Brian F. Bennett Department of Atmospheric and Oceanic Science, University of Maryland, College Park, MD, USA

Timothy P. Canty Department of Atmospheric and Oceanic Science, University of Maryland, College Park, MD, USA

Austin P. Hope Department of Atmospheric and Oceanic Science, University of Maryland, College Park, MD, USA

Ross J. Salawitch Department of Atmospheric and Oceanic Science, University of Maryland, College Park, MD, USA

Department of Chemistry and Biochemistry, University of Maryland, College Park, MD, USA

Earth System Science Interdisciplinary Center, University of Maryland, College Park, MD, USA

Walter R. Tribett Department of Atmospheric and Oceanic Science, University of Maryland, College Park, MD, USA

Chapter 1
Earth's Climate System

Ross J. Salawitch, Brian F. Bennett, Austin P. Hope,
Walter R. Tribett, and Timothy P. Canty

Abstract This chapter provides an overview of the factors that influence Earth's climate. The relation between reconstructions of global mean surface temperature and estimates of atmospheric carbon dioxide (CO_2) over the past 500 million years is first described. Vast variations in climate on geologic time scales, driven by natural fluctuations of CO_2, are readily apparent. We then shift attention to the time period 1765 to present, known as the Anthropocene, during which human activity has strongly influenced atmospheric CO_2, other greenhouse gases (GHGs), and Earth's climate. Two mathematical concepts essential for quantitative understanding of climate change, radiative forcing and global warming potential, are described. Next, fingerprints of the impact of human activity on rising temperature and the abundance of various GHGs over the course of the Anthropocene are presented. We conclude by showing Earth is in the midst of a remarkable transformation. In the past, radiative forcing of climate represented a balance between warming due to rising GHGs and cooling due to the presence of suspended particles (aerosols) in the troposphere. There presently exists considerable uncertainty in the actual magnitude of radiative forcing of climate due to tropospheric aerosols, which has important consequences for our understanding of the climate system. In the future, climate will be driven mainly by GHG warming because aerosol precursors are being effectively removed from pollution sources, due to air quality legislation enacted in response to public health concerns.

Keywords Paleoclimate • Anthropocene • Global warming • Greenhouse gases • Radiative forcing

1.1 Earth's Climate History

Reconstructions of Earth's climate provide a remarkable record of environmental change over vast periods of time. The co-evolution of climate and life on Earth is well established (Schneider 1984; Kasting and Siefert 2002; Sagan and Mullen 1972; Petit et al. 1999). Earth's paleoclimate record is examined here, in some detail, because knowledge of the past is key to understanding the future.

The earliest evidence for life on Earth dates to about 3.5 billion years before present (Bybp) (Brasier et al. 2002). Early life consisted of prokaryotes, one celled bacteria that

© The Author(s) 2017
R.J. Salawitch et al., *Paris Climate Agreement: Beacon of Hope*,
Springer Climate, DOI 10.1007/978-3-319-46939-3_1

thrived in an oxygen (O_2) free environment. These organisms had no nucleus and reproduced by cell division. The first prokaryotes likely made organic matter by combining carbon dioxide (CO_2) with molecules such as hydrogen sulfide (H_2S), releasing water (H_2O) and elemental sulfur to the environment (Canfield and Raiswell 1999).

Early in Earth's history the favored atmospheric fate for carbon-bearing compounds was methane (CH_4), because the atmosphere was in a state chemists call "reducing". Stellar astronomy indicates that at the time early life formed, the luminosity of our Sun was about 30 % less than today, which should have caused ancient oceans to freeze. As explored in detail throughout this book, CH_4 is a more potent greenhouse gas (GHG) than CO_2. Extremely high levels of atmospheric CH_4 and ammonia (NH_3), another reduced compound that is also a strong GHG, were likely responsible for preventing Earth's ancient oceans from freezing (Sagan and Mullen 1972).

Prokaryotes were the first to develop photosynthesis, the ability to convert sunlight, carbon dioxide (CO_2), and water (H_2O) into glucose $C_6H_{12}O_6$. Eventually, prokaryotic photosynthesis caused atmospheric O_2 to rise from about one part per million of all air molecules to 21 %. Margulis and Sagan (1986) call the initial build-up of atmospheric O_2 the greatest environmental crisis Earth has ever endured. At the time, O_2 was toxic to most life on Earth. As a result, a mass extinction called the Great Oxygenation Event occurred about 2.5 Bybp. One can only imagine the emergency meetings of bacterial communities, seeking to ban their photosynthetic cousins in an effort to halt the build-up of atmospheric O_2.

The rise of atmospheric O_2 had enormous consequences. For the first time in Earth's history, CO_2 was the favored state for atmospheric carbon gases. Conversion of atmospheric CH_4 to CO_2 likely led to Earth's first glaciation event about 2.4 Bybp (Frei et al. 2009). The build-up of O_2 also led to formation of Earth's protective ozone (O_3) layer, which was necessary for life to emerge from sea to land. Finally, the global, atmospheric chemical shock induced by the Great Oxygenation Event facilitated the evolution of eukaryotes: nucleated cells that metabolize O_2. You are made of eukaryotic cells!

Plant life first appeared on land about 500 million years before present (Mybp) (Kenrick and Crane 1997). Even though, as alluded to above, much is known about climate and the state of Earth's atmosphere prior to this time, reconstructions of global variations in Earth's climate and atmospheric CO_2 are only available for the most recent 500 million years.

Figure 1.1 shows the variations in the global mean surface temperature and the abundance of carbon dioxide (CO_2) over the past 500 million years. The temperature estimates are anomalies (ΔT) with respect to the mean state of Earth's climate that existed during recent pre-industrial time (i.e., years 1850–1900). Notable events regarding the evolution of life (Dinosaurs, Rise of Mammals, etc.) are indicated to the left and phenomenon regarding the global carbon cycle and climate (Rise of Forests, Greenland Glaciation, etc.) are marked to the right. This figure is our composite of a considerable number of paleoclimatic studies, as described in Methods. To facilitate discussion of this figure, six Eras are denoted. These should not be confused with the formal use of the word Era by Geologists. Each era in Fig. 1.1 spans a different length of time; the interval over which Dinosaurs lived (about 230–65 Mybp) is about 2.5 times longer than the time between the Rise of Mammals and present.

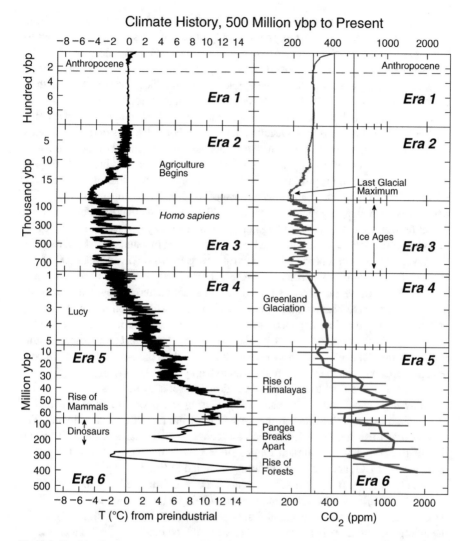

Fig. 1.1 Earth's climate history, past 500 million years. Historical evolution of global mean sur-face temperature anomaly (ΔT) relative to a pre-industrial baseline (i.e., mean value of global temperate over 1850–1900) (*left*) and the atmospheric mixing ratio of CO_2 (*right*). Major events in the evolution of life on Earth as well as either changes in climate of the global carbon cycle are denoted. The *vertical line* on the ΔT panel at zero marks the pre-industrial baseline; the *vertical lines* on the CO_2 panel denote mixing ratios of 280 ppm (pre-industrial), 400 ppm (current level) and 560 ppm (twice pre-industrial level). These time series are based on hundreds of studies; see Methods for further information

Figure 1.1 shows that Earth has undergone vast changes in climate as well as the abundance of atmospheric CO_2. A logarithmic scale is used to represent the mixing ratio of atmospheric CO_2, because the radiative forcing of climate (see Sect. 1.2.1) is proportional to the logarithm of CO_2. Figure 1.1 shows, clearly and beyond debate, the strong association of Earth's climate and atmospheric CO_2. Looking backwards in time, as CO_2 rises, Earth warms.

On geological time scales, atmospheric CO_2 is controlled by the carbonate-silicate cycle (Berner et al. 1983). Atmospheric supply of CO_2 occurs during volcanic eruptions and hydrothermal venting. Atmospheric removal of CO_2 is more complicated. The weathering of minerals converts atmospheric CO_2 into a water soluble form of carbon; ocean organisms incorporate soluble carbon into their shells and, when these animals perish, their shells sink to the ocean floor. Plate tectonics buries the sinking sediment, after which the carbon either remains in Earth's mantle or, on occasion, is spewed back to the atmosphere-ocean system, via either volcanoes or deep sea vents.

The first dramatic perturbation to the carbonate-silicate cycle was induced by the rise of forests. About 500 Mybp, atmospheric CO_2 may have been as high as 5000 parts per million by volume (ppm), more than a factor of 10 larger than today (Fig. 1.1) (Berner 1997). Of course, Earth was also exceedingly warm compared to today. The first plants to evolve, bryophytes, were algae-like organisms that probably eased the transition from sea to land by finding homes on moist rocky surfaces. Bryophytes are known as non-vascular plants; they lack roots to transport moisture. Moss is a modern-day bryophyte. Vascular pteridophytes (fern-like organisms) evolved about 500 Mybp, soon leading to the rise of forests. This resulted in a steady, dramatic decline in atmospheric CO_2 because early forests lacked the abundant bacteria, fungi, and small soil animals that recycle plant matter in contemporary forests. Carbon in the forests that prevailed during the time depicted as Era 6 of Fig. 1.1 was buried and converted to modern day coal and natural gas deposits, due to the intense heat and pressure within Earth's mantle. This carbon is now being released back to the atmosphere–ocean system, perhaps to generate the electricity used to help you read this book.

The next event that transformed the global carbon cycle was the rise of the Himalayas (Raymo and Ruddiman 1992). During the period of time depicted in Era 5, plate tectonics resulted in the formation of the modern-day continents. The exposure of fresh minerals due to the vast tectonic activity associated with formation of the Himalayan mountain range, the largest in the world, led to the steady draw down of CO_2 and associated cooling depicted in Era 5 of Fig. 1.1.

About 3 Mybp, two remarkable events occurred. Our predecessor Lucy (*Australopithecus afarensis*) roamed modern day Ethiopia (Johanson and White 1979). At about the same time, Greenland first became glaciated (Lunt et al. 2008). While the emergence of an early human ancestor who walked in an upright manner is in no way related to the glaciation of Greenland, it is worth noting that global mean surface temperature and atmospheric CO_2 at the time of Lucy were both estimated to be at modern, pre-industrial levels (Era 4, Fig. 1.1).

The most compelling association of CO_2 and climate is provided by the co-variance of these quantities during the past 800,000 years (Era 3, Fig. 1.1) (Imbrie and Imbrie 1979). Earth's climate oscillated between glaciated and inter-glacial states, with atmospheric CO_2 levels of about 200 ppm and 280 ppm, respectively, characterizing each state (Barnola et al. 1987). Lower levels of atmospheric CO_2 prevailed during glacial times due to more productive ocean biogeochemical uptake (Marino et al. 1992), perhaps facilitated by the oceanic supply of iron resulting from the grinding of glaciers on rock (Martin 1990). The ultimate pace-maker of these cycles is orbital variations of Earth about the Sun, known as Milankovitch cycles (Imbrie and Imbrie 1979).

The precise timing of the rise and fall of temperature and CO_2 during Era 3 of Fig. 1.1 is a source of considerable dispute between the climate "believers" and "deniers". The initial, literal interpretation of the ice core record suggested that changes in temperature proceeded variations in CO_2 by about 800 years (Caillon et al. 2003). If so, the deniers argue, then CO_2 is responding to, rather than driving, global climate change. It is essential to appreciate that: the ice core record of CO_2 is discerned by measuring the composition of bubbles trapped in ice; historic temperature is quantified by measuring isotopic composition of the hydrogen and/or oxygen elements within the ice; and bubbles within the sampled ice cores move with respect to the surrounding ice over geologic time. A recent re-analysis of the timing of variations in temperature and CO_2 of an Antarctic ice core, which considers movement of bubbles with respect to the surrounding ice, reveals synchronous variation within the uncertainty of measurement (Parrenin et al. 2013). This interpretation supports the view that changes in atmospheric CO_2 did indeed drive glacial/interglacial transitions.

Ancient air preserved in ice cores reveals that when Earth underwent glacial conditions during Era 3 of Fig. 1.1, atmospheric CH_4 fell to about 0.4 ppm. During interglacial periods, atmospheric CH_4 reached a value of 0.7 ppm (Petit et al. 1999; Loulergue et al. 2008). Natural sources of CH_4 vary by an amount large enough to induce considerable variations in atmospheric abundance, with some consequence for the radiative forcing of climate. Methane is released to the atmosphere when frequently flooded regions (wetlands) experience low oxygen (anaerobic) conditions. It is likely that the higher levels of CH_4 during warm epochs was due to a larger preponderance of wetlands, particularly in the northern hemisphere (NH), as these regions went from ice-covered to ice-free as Earth transitioned from glacial to interglacial conditions (Brook et al. 2000; Sowers 2006). Variations in atmospheric CH_4 also played a role in driving glacial/interglacial climate cycles.

The correlation of temperature and atmospheric CO_2 over vast periods of Earth's history is firmly established by the hundreds of studies that have led to our composite Fig. 1.1. Of course, correlation does not necessarily imply causation. The radiative forcing of climate due to CO_2 and other GHGs is explored in great detail later in this chapter.

Modern Homo sapiens evolved about 200,000 years ago and left Africa about 100,000 years ago (Carto et al. 2009). The paths of early humans were influenced by various rapid climate change events that took place at the end of Era 3 and the start of Era 2 of Fig. 1.1. During the height of most recent glaciation, about 20,000 years ago, modern day Manhattan was under a sheet of ice nearly half a mile thick and global sea level was 120–130 m (about 400 ft!) lower than today. Human settlement of North America hugged the coastline, as the interior was inhospitable if not impassable. Scandinavia, England, Wales, Scotland, and Ireland were similarly buried under year-long ice.

Modern agriculture was invented during the Neolithic Revolution, about 12,500 years ago. The Earth was in the midst of climatic warming following the end of the last ice age, which has been implicated as a causal factor because the domestication of plants and animals occurred nearly simultaneously at places separated by great distance (Gepts and Papa 2001). Agriculture flourished, population grew, and humans colonized all parts of the Earth (except Antarctica) during the climatically quiescent times depicted at the end of Era 2 and start of Era 1 in Fig. 1.1.

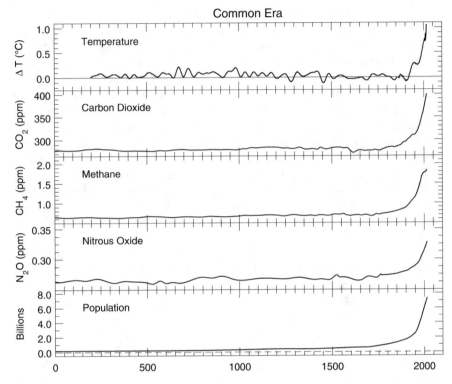

Fig. 1.2 Temperature, GHGs, and population, Common Era. Time series of Earth's global mean surface temperature anomaly (ΔT) relative to pre-industrial baseline (1850–1900 mean) (Jones and Mann 2004; Jones et al. 2012), the atmospheric mixing ratio of CO_2, CH_4, and N_2O (MacFarling Meure et al. 2006; Ballantyne et al. 2012; Dlugokencky et al. 2009; Montzka et al. 2011) and global population (Klein Goldewijk et al. 2010; United Nations 2015) over the Common Era. See Methods for further information

Eventually, population rose to an extent that humans began to exert a measurable effect on the GHG levels of the atmosphere, a time period now known as the Anthropocene (Crutzen and Stoermer 2000). Figure 1.2 illustrates the time evolution of ΔT, CO_2, CH_4, nitrous oxide (N_2O), and population during the past 2000 years, the so-called Common Era. The association of the rise in population and increased atmospheric levels of CO_2, CH_4, and N_2O is again irrefutable. Before delving into the Anthropocene, we shall comment on one more controversy.

Europe experienced unusually warm temperatures from about 950–1250 AD, a time known as the Medieval Warm Period (MWP) (Moberg et al. 2005). Our Fig. 1.2, which relies on the ***global*** temperature reconstruction of Jones and Mann (2004), does not depict the MWP. The study of Jones and Mann (2004) suggests the MWP was regional in nature, with little to no global expression. The temperature record in Fig. 1.2, which has become known as "The Hockey Stick", has led to considerable controversy. One account is described in the book *The Hockey Stick and the Climate Wars: Dispatches from the Front Lines* (Mann 2012). In 2006, the

United States National Academy of Sciences (NAS 2006) reviewed the voluminous literature on climate reconstructions over the Common Era and concluded:

> Based on the analyses presented in the original papers by Mann et al. and this newer supporting evidence, the committee finds it plausible that the Northern Hemisphere was warmer during the last few decades of the 20th century than during any comparable period over the preceding millennium. The substantial uncertainties currently present in the quantitative assessment of large-scale surface temperature changes prior to about A.D. 1600 lower our confidence in this conclusion compared to the high level of confidence we place in the Little Ice Age cooling and 20th century warming. Even less confidence can be placed in the original conclusions by Mann et al. (1999) that 'the 1990s are likely the warmest decade, and 1998 the warmest year, in at least a millennium' because the uncertainties inherent in temperature reconstructions for individual years and decades are larger than those for longer time periods and because not all of the available proxies record temperature information on such short timescales.

This NAS statement provides fodder for both the believers and deniers. The deniers posit that if global temperature was indeed unusually warm from 950 to 1250 AD, at a time when CO_2, CH_4, and N_2O were known to be stable (Fig. 1.2), then other factors such as solar luminosity must be responsible. If so, the argument goes, then perhaps the late twentieth century and early twenty-first century warming is due to some factor other than anthropogenic GHGs.

A recent study of ocean heat content supports the view that the higher temperature of the MWP was indeed global in nature (Rosenthal et al. 2013), similar to the conclusion reached by Soon and Baliunas (2003) a decade earlier. If so, then the time series of ΔT shown in Fig. 1.2 is in need of revision. Regardless, as shown below, extremely strong scientific evidence implicates GHGs produced by human activities as the primary driver of rising global temperature during the past half-century.

1.2 The Anthropocene

The Anthropocene refers to the recent interval during which the atmospheric abundance of GHGs that drive Earth's climate have increased due to human activity (Crutzen and Stoermer 2000). Most peg the start of the Anthropocene to the mid-eighteenth century, linked to the invention of the steam engine by James Watt in 1784 (Steffen et al. 2015). Others suggest humans have had a discernable influence on GHGs for a much longer period of time, and argue for a start date to the Anthropocene as far back as 8000 years before present (ybp) (Ruddiman 2003).

We use 1765 as the start of the Anthropocene for several reasons. The largest influence of humans on GHGs has certainly occurred since 1765. Estimates of radiative forcing of climate (defined in Sect. 1.2.1) due to a wide variety of human activities based on a multi-year effort of scientists from many nations, are available in a transparent, easily accessible format[1] back to 1765 (Meinshausen et al. 2011).

---

[1] RF estimates in ASCII and Excel format are available at:

http://www.pik-potsdam.de/~mmalte/rcps/data/20THCENTURY_MIDYEAR_RADFORCING.DAT

http://www.pik-potsdam.de/~mmalte/rcps/data/20THCENTURY_MIDYEAR_RADFORCING.xls

Finally, changes in global mean surface temperature, to which RF of climate will be related, are much more certain from 1765 to present than for times extending back to the invention of agriculture (NAS 2006). Our choice is not meant to dismiss the importance of human influence on climate prior to 1765. If, as suggested by Ruddiman (2003), human activity 8000 ybp did indeed offset the onset of extensive glaciation due to declining summer insolation at Northern high latitudes (driven by Milankovitch orbital variations), this would be a fascinating benefit of human ingenuity, especially for indigenous peoples of high northerly latitudes.

Rather than wade deeper into the debate over the start of the Anthropocene, we next describe a few figures that illustrate the human fingerprint on the global carbon cycle and climate change over the past several centuries. Along the way, the mathematical principles needed to understand the material presented in Chaps. 2 and 3 are developed.

1.2.1 Radiative Forcing

In the absence of an atmosphere, the temperature of Earth would be governed by:

$$T_{EARTH} = \left(\frac{(1 - Albedo)\,S/4}{\sigma} \right)^{\frac{1}{4}} \tag{1.1}$$

Albedo, the Latin word for whiteness, refers to the fraction of incoming sunlight reflected to space (commonly about 0.3 (or 30%) for Earth), S is the luminosity of our Sun at the distance of Earth's orbit (1370 W m^{-2}), and σ is the Stefan Boltzmann constant (5.67$\times$10^{-8} W m^{-2} K^{-4}). A value of $S/4$ is used because the Earth intercepts sunlight like a disk and radiates heat like a sphere; 1/4 is the ratio of the surface area of a disk to that of a sphere (this concept as well as Eq. 1.1 are explained in many introductory Earth Science textbooks). Solving for the putative temperature of planet Earth without an atmosphere yields 255 K, which is −18° Celsius (°C) or 0° Fahrenheit (°F). This is much colder than the average temperature of today's Earth. If the temperature found using Eq. 1.1 actually applied, our oceans would be frozen.

The greenhouse effect, the trapping of radiation by our planet's atmosphere, is responsible for the difference between the Earth's actual temperature and that found using Eq. 1.1. Earth's mean surface temperature is about 15.5 °C or 60 °F. Earth's atmosphere is responsible for increasing the amount of energy the surface receives, by several hundred W m^{-2}, in comparison to an Earth devoid of an atmosphere. This excess heat is driven by the abundance and molecular properties of GHGs such as H_2O, CO_2, CH_4, and N_2O, as well as clouds (i.e., condensed H_2O droplets).

Infrared radiation (or heat in the form of photons) emitted by Earth's surface is resonant with various vibrational modes of GHG molecules, inducing these photons to be absorbed and re-emitted in all directions. Some of this absorbed and re-emitted radiation is sent back to the surface. As such, GHGs in Earth's atmosphere act as a blanket, trapping heat that would otherwise escape to space. Water vapor, the most important GHG, is responsible for the majority of radiation sent back to the surface. The abundance of H_2O in our atmosphere is controlled by thermodynamics: i.e., the evaporation of H_2O from the oceans, condensation of H_2O in the atmosphere to form clouds, and deposition of H_2O back to the surface in the form of precipitation. Atmospheric H_2O varied prior to the onset of human influence; the effect of thermodynamics on various isotopes of H_2O preserved in ice cores is an important tool for quantitative reconstruction of past climate (Jouzel et al. 1987).

The radiative forcing (RF) of climate refers to the increase in the amount of heat directed to Earth's lower atmosphere as the abundance of GHGs rise. Here and throughout this book, we follow the convention established in the 2001 Intergovernmental Panel on Climate Change (IPCC) Physical Science Basic Report (IPCC 2001) that RF of climate is defined as the change in the net flow of energy (sunlight plus infrared heat) at the tropopause (boundary between the lower atmosphere, or troposphere, and the upper atmosphere, or stratosphere) relative to a particular start date, after allowing for stratospheric temperatures to adjust to radiative equilibrium. This concept is explained well in Sect. 2.2 of IPCC (2007).

Figure 1.3a shows several time series of the RF of climate over the Anthropocene. The curves are set to zero in year 1765 and represent changes relative to this start time,

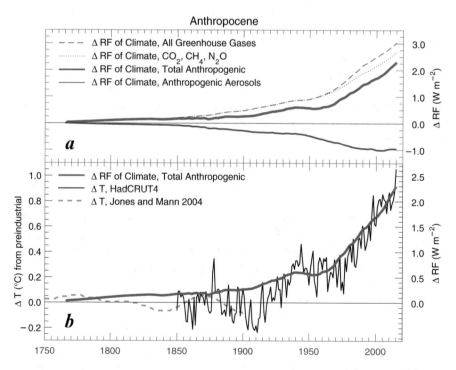

Fig. 1.3 Radiative forcing of climate and temperature, Anthropocene. (**a**) change in the radiative forcing of climate (ΔRF) relative to year 1765, from various factors (as indicated) for the RCP 4.5 scenario (Meinshausen et al. 2011); (**b**) total anthropogenic ΔRF (*red*) and the global mean surface temperature anomaly (ΔT) relative to pre-industrial baseline (1850–1900 mean) from the modern instrument record (HadCRUT4) (Jones et al. 2012) and from various proxies (Jones and Mann 2004). See Methods for further information

hence they are denoted ΔRF.[2] Red is used to represent warming (positive ΔRF); blue is used to show cooling (negative ΔRF). These ΔRF curves are based on the GHG and aerosol precursor abundances used to drive climate model simulations of IPCC (2013). The RF of climate due to human release of CO_2, CH_4, and N_2O is the primary focus of this book. The historic ΔRF due to all anthropogenic GHGs (dashed red) exceeds that of the $CO_2/CH_4/N_2O$ triplet (dotted red) by a small amount, with most of the difference due to a class of compounds called Ozone Depleting Substances (ODS). Even though ODS exhibit a greenhouse effect, they are generally not labeled as GHGs because their most important detrimental effect is depletion of Earth's ozone layer. Also, industrial production of ODS has been successfully curtailed by the Montreal Protocol and the effect of these compounds on climate will diminish in the future (WMO 2014).

The human release of pollutants that increase the burden of small particles in the troposphere, known as aerosols, leads to a reduction in the RF of climate (blue line, Fig. 1.3a). This occurs because many aerosols reflect sunlight. An estimate of ΔRF due to aerosols provided by Meinshausen et al. (2011) is shown in Fig. 1.3a.[3] As detailed below, there is considerable uncertainty in this term.

The total ΔRF due to human activity is shown by the solid red curve in Fig. 1.3a, b. All told, human activities have increased the RF of climate by about 2.3 W m^{-2} between 1750 and present. Figure 1.3b shows that the time variation of total ΔRF due to humans (red line) closely resembles the observed rise in global mean surface temperature anomaly (black and grey lines). Below we conduct quantitative analysis of these two quantities, both within this chapter as well as throughout Chap. 2.

Figure 1.4 details the change in RF of climate, between 1750 and 2011, due to various factors.[4] Numerical values are from Chap. 8 of IPCC (2013). Error bars denote the 5 and 95 % confidence intervals for each quantity. Processes that effect RF of climate but are not regulated under GHG legislation, such as Stratospheric and Tropospheric Ozone and Land Use Change, as well as minor terms such as Contrails and Solar Irradiance, are also shown. The solid red lines of Fig. 1.3 consider all of the terms shown in Fig. 1.4.

There are several aspects of Fig. 1.4 worth emphasizing. Human release of GHGs has warmed climate, with CO_2 being the most important contributor. The combined effect of the two other most important long-lived anthropogenic GHGs (CH_4, N_2O) plus all of the ODS compounds has enhanced this CO_2-based warming by about 60 %. Tropospheric ozone (O_3), a pollutant harmful to human health and agriculture, has warmed climate over the course of the Anthropocene by nearly as much as CH_4. Tropospheric O_3 is regulated by air quality regulations that vary by country and focus

---

[2] Delta is the first letter of the Greek word diaphorá, which means difference. Scientists often use either Δ (capital delta) or δ (lowercase delta) to represent difference. The ΔRF data in Fig. 1.3 start in 1765 because this is the first year for which numerical values are available (see Methods).

[3] The blue line is the combination of the three terms: the direct radiative effect of aerosols, the perturbation to the reflectivity of clouds induced by aerosols, and the darkening of snow caused by the deposition of black carbon. This estimate includes the following types of aerosols: sulfate, nitrate, mineral dust, as well as organic carbon and black carbon from both fossil fuel combustion and biomass burning.

[4] Figure 1.3 spans 1765–2011 whereas Fig. 1.4 tabulates ΔRF between 1750 and 2011. The different start years are a result of how scientists who worked on IPCC (2013) handled various data streams. This difference is inconsequential since human activity imposed very little influence on ΔRF between 1750 and 1765.

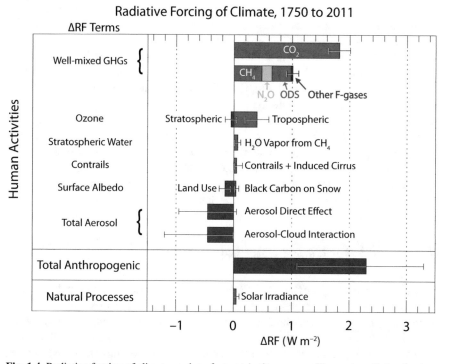

Fig. 1.4 Radiative forcing of climate, various factors, Anthropocene. Change in radiative forcing of climate (ΔRF) over the course of the Anthropocene (in this case, 1750–2011) due to human factors (GHGs and aerosols) and natural processes (solar irradiance). Error bars represent the 5–95 % confidence interval. The ODS entry represents ΔRF due to ozone depleting substances such as CFC-11, CFC-12, etc. The Other F-gases entry represents ΔRF due to HFCs, PFCs, SF_6, and NF_3. Numerical estimates for ΔRF are shown when available; otherwise, numerical estimates for the change in Effective Radiative Forcing (ΔERF) are used. After Fig. 8.15 of IPCC (2013). See Methods for further information

on surface conditions (i.e., the air we breathe). There is no coordinated international effort to limit future growth of tropospheric O_3. Human release of pollutants that lead to formation of tropospheric aerosols causes climate to cool due to two processes nearly equal in magnitude but having large uncertainty: the reflection of sunlight by aerosols (Aerosol Direct Effect) and the effect of aerosols on cloud formation (Aerosol-Cloud Interaction).[5] The net effect of all human activity on ΔRF (bar labeled Total Anthropogenic) is about 25 % larger than ΔRF due to CO_2 and has a considerably larger uncertainty (size of respective error bars) (Chap. 8, IPCC (2013)).

Other aspects of global warming are represented in Fig. 1.4. The reflectivity of Earth's surface has changed primarily due to deforestation that makes the surface

---

[5] The effect of anthropogenic aerosols on the radiative properties of clouds is different, from a climate modeling perspective, than the evolution of cloud properties as the surface warms. Specialists refer to the former as the aerosol indirect effect and the latter as cloud feedback. A considerable research effort informs us that the aerosol indirect effect leads to a cooling (negative RF) of uncertain magnitude and the cloud feedback could either lead to warming or cooling.

brighter (leading to cooling) and the deposition of dark, carbonaceous material on snow that darkens the surface (leading to warming). Contrails from aircraft have led to a slight warming, mainly because of induced clouds. A slight warming term is also attributed to an increase in stratospheric humidity driven by rising tropospheric CH_4, which is converted to H_2O when lost in the stratosphere. The depletion of stratospheric O_3 has resulted in slight cooling. Finally and most importantly, scientists have been able to estimate the time variation of total solar irradiance over the course of the Anthropocene. The trend in solar irradiance over this two and a half century time period is small. Even if the MWP discussed above turns out to be as warm as the present decades, presumably due to an increase in solar irradiance during the middle of the Common Era (Bard et al. 2000), scientists are nonetheless confident the rise in temperature over the Anthropocene was driven by rising GHGs and not a change in the luminosity of our Sun (Chap. 8, IPCC (2013)).

1.2.2   Global Warming Potential

The global warming potential (GWP) metric was developed to guide public policy decisions regarding trade-offs of release of various GHGs. The GWP of a particular compound represents the ratio of the rise in global mean surface temperature (GMST) due to the release of a particular amount (mass) of this compound, relative to the rise in GMST resulting from the release of the same amount (mass) of CO_2. Inherent in the computation of GWP is that the increase in GMST is found over a particular time horizon.

The most commonly used time horizons are 20 and 100 years. The mathematical expression for the GWP of CH_4 over a 100-year time horizon, which relies on the use of a calculus concept called integration, is given by:

$$GWP(CH_4) = \frac{\int_0^{100\,yrs} a_{CH_4} \times CH_4(t)\,dt}{\int_0^{100\,yrs} a_{CO_2} \times CO_2(t)\,dt} \qquad (1.2)$$

where a_{CH_4} and a_{CO_2} are the radiative efficiencies (units W m^{-2} kg^{-1}) of CH_4 and CO_2, respectively, and $CH_4(t)$ and $CO_2(t)$ represent the time dependent response to the release into the atmosphere of the same mass of these two GHGs.[6] The atmospheric lifetime of a GHG denotes the time it takes for a pulsed release of the gas to decay by $1/e$ of the initial value, where e ≈ 2.718. The lifetimes of CH_4

---

[6] Equation 1.2 represents a computer simulation of the cumulative radiative forcing of climate over a 100 year time period due to release of a pulse of a certain amount (mass) of CH_4, divided by the cumulative radiative forcing over the same time period due to simulated release of the same amount of CO_2. Since the pulse of CH_4 decays faster than the pulse of CO_2, due to the shorter lifetime of CH_4, the GWP of CH_4 is larger when shorter time periods are considered.

Table 1.1 Global warming potentials

GHG	IPCC (1995)	IPCC (2001)	IPCC (2007)	IPCC (2013)
100-Year time horizon				
CH_4	21	23	25	28, 34[a]
N_2O	310	296	298	265, 298[a]
20-Year time horizon				
CH_4	56	62	72	84, 86[a]
N_2O	280	275	289	264, 268[a]

[a]Allowing for carbon cycle feedback

and N_2O used in IPCC (2013) calculation of GWP are 12.4 and 121 years, respectively. These lifetimes are typically used for evaluation of the numerator of Eq. 1.2, as most GWP estimates assume pure exponential decay. Conversely, the decay of the pulse of CO_2 in the denominator of Eq. 1.2 is found using a computer model of the global carbon cycle.

As shown in Chap. 3, the GWP of various GHGs is vitally important for assessing the efficacy of the Paris Climate Agreement[7]. Table 1.1 provides the GWP of CH_4 and N_2O from the past 4 IPCC reports. The GWP of GHGs has been updated over time due to evolving knowledge of the radiative efficiencies and lifetimes of atmospheric compounds. Also, IPCC (2013) provided two values of GWP for CH_4 and N_2O: with and without consideration of carbon cycle feedback.

The most commonly used GWPs for public policy are those found for a 100-year time horizon. This preference is traceable to the 1997 Kyoto Protocol of the United Nations Framework Convention on Climate Change (UNFCCC), which was based on 100-year GWPs. Furthermore, since the values of GWP given in IPCC (2013) that do not allow for carbon cycle feedback are most analogous to values of GWPs given in prior IPCC reports, we will use GWPs for CH_4 and N_2O of 28 and 265, respectively, in our analysis of the Paris Climate Agreement. The GWPs of other GHGs used in our analysis are based on Chap. 8 of IPCC (2013).

The fact that the GWPs of CH_4 and N_2O are much larger than 1 means that, on a per mass basis, these GHGs are considerably more potent than CO_2.[8] Furthermore, the GWP of CH_4 is much larger over a 20-year horizon than a 100-year time horizon, due to the 12.4 year lifetime for CH_4 used in the calculation of GWP. Integrated over 20

---

[7]The Paris Climate Agreement was negotiated at the 21st Conference of the Parties (COP21) held in Paris, France during December 2015. The COP meetings are an annual gathering of representatives from participating nations, environmental agencies, and industry to address concerns of climate change. For more information see: http://unfccc.int/paris_agreement/items/9485.php

[8]Some textbooks and reports provide GWP values on a per molecule basis, rather than a per mass basis. A molecule of CO_2 with atomic mass of 44 weighs 2.75 times a molecule of CH_4 (atomic mass of 16). Using the IPCC (2013) value for the GWP of CH_4 on a 100 year time horizon, without consideration of carbon cycle feedback, scientists would state CH_4 is 28 times more potent than CO_2 on a *per mass* basis and, at the same time, is 10.2 ($28 \div 2.75$) times more potent than CO_2 on a *per molecule* basis.

years, a significant fraction of the initial, pulsed release of CH_4 is present in the modeled atmosphere. However, integrated over 100 years, a much smaller fraction is present. As discussed in Chap. 4, controlling inadvertent release of CH_4 to the atmosphere will likely be vitally important for reaching the goals of the Paris Climate Agreement, keeping warming well below 2.0 °C and aiming to limit warming to 1.5 °C. The importance of CH_4 would be amplified by a factor of 3 (ratio of 84/28 from Table 1.1) if climate change over a 20-year time horizon were used to guide public policy, placing even more stringent controls on the atmospheric release of this GHG.

Carbon dioxide is the most important anthropogenic GHG for RF of climate (Figs. 1.3a and 1.4), despite the more potent nature of CH_4 and N_2O, because human society has released to the atmosphere a much greater mass of CO_2 than other GHGs. Simply put, CO_2 is the greatest waste-product of modern society. We now turn our attention towards the human fingerprints on global warming as well as on the atmospheric build-up of CO_2 and other GHGs.

1.2.3 Human Fingerprints

As described in Sect. 1.1, Earth's climate has undergone vast variations on geologic time scales. Many of these climate shifts are directly tied to changes in atmospheric CO_2. Studies of paleoclimate must also consider effects on global mean surface temperature (GMST) of continental plate alignment (Hay et al. 1990), the seasonal distribution of sunlight related to variations of Earth's orbit (Erb et al. 2013), as well as the radiative forcing of climate due to aerosols (Chylek and Lohmann 2008) and GHGs other than CO_2 (Sagan and Mullen 1972).

Even though Earth's climate and the abundance of GHGs co-vary on geologic time scales due to natural processes, a scientific consensus has nonetheless emerged that the recent rise in GMST as well as atmospheric burdens of CO_2, CH_4, and N_2O are all driven by human activities (IPCC 2013). Here we briefly review some of the most important human fingerprints on temperature and GHGs. To place the material that follows in the proper perspective, it is important to understand that the time scales involved with geologic change and human history are enormously different. For instance, the ratio of the time since the rise of forests (400 Mypb) to the time since the advent of agriculture (~12,000 ybp) is enormous. If time on Earth since the rise of forests were compressed into a 24 h day, the time since the advent of agriculture would take 2.6 s, which is less than the time it takes to read this sentence!

1.2.3.1 Rising Temperature

As noted above, correlation does not demonstrate causation. We shall first examine the quantitative relation between the rise in the GMST anomaly (ΔT) shown in Fig. 1.3 and the change in radiative forcing of climate (ΔRF) attributed to humans over the Anthropocene.

The Stefan-Boltzmann equation relates the temperature of an object to the rate at which it is able to disperse energy:

$$Power = \sigma T^4 \qquad (1.3)$$

Power is used in Eq. 1.3 since *Power* is defined as *Energy/Time*, and the Stefan-Boltzmann equation is based on the rate at which energy is dispersed; σ is the same constant used in Eq. 1.1. In equilibrium, Earth's surface releases (or radiates) energy at the same rate it is supplied to the surface by the atmosphere. Hence *Power* in Eq. 1.3 can be replaced with RF, where RF represents the atmospheric radiative forcing of climate:

$$RF = \sigma T^4 \qquad (1.4)$$

Those who have taken calculus will understand that upon taking the derivative of Eq. 1.4 and re-arranging terms, it can be shown that:

$$\Delta T = \frac{1}{4\sigma T^3} \Delta RF \qquad (1.5)$$

where ΔRF represents a perturbation to the system (i.e., the rise in RF of climate due to human release of GHGs) and ΔT represents the response (i.e., resulting rise in global mean surface temperature).

Scientists use relations such as Eq. 1.5 to diagnose the output of climate models. A common value for the term $\frac{1}{4\sigma T^3}$ in Eq. 1.5 is 0.31 K/(W m^{-2}), which is related to the temperature at which Earth radiates to space (Bony et al. 2006). Substituting this numerical value for $\frac{1}{4\sigma T^3}$ into Eq. 1.5 leads to:

$$\Delta T = 0.31 \frac{K}{W\,m^{-2}} \Delta RF \qquad (1.6)$$

We now examine the **quantitative consistency** between the rise in temperature (ΔT in Fig. 1.3) over the Anthropocene and the radiative forcing of climate attributed to humans (ΔRF in Fig. 1.3). The product $0.31 K / W\,m^{-2} \times 2.3 W\,m^{-2}$ is equal to 0.7 K (which is the same as 0.7 °C),[9] quite close to the observed rise in GMST (about 0.9 °C) over the course of the Anthropocene. As examined in detail in Chap. 2, the actual relationship between ΔT and ΔRF requires a consideration of factors such as climate feedback (i.e., enhancement or diminution of the RF of climate imposed by humans due to changes in factors such as atmospheric humidity and clouds) as well as the transport of heat from the atmosphere to the world's oceans, which specialists refer to as ocean heat export. It is likely, for instance, that positive feedback (enhancement) of the direct RF of climate caused by humans is

---

[9] Degrees Celsius and degrees Kelvin are identical when used to express temperature difference.

responsible for the difference between the expected (0.7 °C) and observed (0.9 °C) rise in GMST over the course of the Anthropocene. Most importantly, the calculation conducted above reveals *quantitative consistency* between the observed and expected rise in global temperature over the course of the Anthropocene. This is a critically important first step in the attribution of global warming to humans.

Several other aspects of global warming bear the human fingerprint. Climate models predict that as GHGs rise, the lowest layer of the atmosphere (the troposphere) will warm while the second layer of the atmosphere (the stratosphere) should cool. Stratospheric cooling is a consequence of the blanketing effect of GHGs: as atmospheric levels of GHGs rise, a larger fraction of the thermal energy emitted by the surface is absorbed, re-emitted, and therefore blocked from reaching the stratosphere. As shown in Fig. 1.5, tropospheric warming coupled with stratospheric cooling is seen in the climate record, at least over the part of the Anthropocene for which modern measurements of atmospheric temperature profiles exist (Sherwood and Nishant 2015). About two-thirds of the cooling of the upper stratosphere for the time period 1979–2005 has been attributed to rising GHGs, with the remainder attributed to human-induced depletion of stratospheric O_3 (Mitchell 2016). The pattern of temperature changes with respect to altitude and latitude throughout the troposphere and stratosphere agrees with the pattern predicted by climate models to a high degree of statistical significance (Santer et al. 2013a), although these models do tend to overestimate the amount of warming observed in the lower atmosphere (Santer et al. 2013b).[10]

Another important human fingerprint of global warming is the observation that the altitude of the tropopause, the boundary between the troposphere and the stratosphere, has been rising as predicted by climate models (Santer et al. 2013a). Had modern global warming been due to an increase in the luminosity of the Sun or a release of energy from the world's oceans, scientists would have expected to observe warming in the stratosphere and troposphere as well as little to no change in the height of the tropopause.

1.2.3.2   Carbon Dioxide

Carbon dioxide (CO_2) is the single greatest waste product of modern society. There is compelling scientific evidence that the rise in atmospheric CO_2 during the Anthropocene is due, nearly entirely, to human activity. The rise in CO_2 from 1765 to the early 1900s was predominately driven by the clearing of forests for agriculture (also known as land use change, or LUC) (Siegenthaler and Oeschger 1987). For a few decades subsequent to 1900, LUC and the combustion of fossil fuels made nearly equal contributions to the rise in atmospheric CO_2. Since the early 1950s, the growth of atmospheric CO_2 has been driven primarily by the combustion of fossil fuels[11] (Le Quéré et al. 2015).

---

[10]The tendency of many climate models to overestimate observed warming is a central theme of Chap. 2.

[11]Combustion of fossil fuels refers to the burning of coal, oil and gasoline, as well as natural gas (mainly methane) to meet society's needs for heat, electricity, transportation, and various other industrial enterprises.

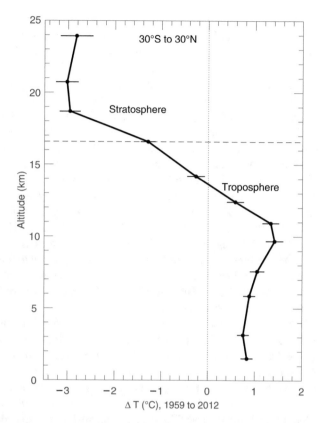

Fig. 1.5 Temperature change profile, 1959–2012. Temperature difference, 1959–2012, based on analysis of radiosonde observations acquired between latitudes of 30°S and 30°N (positive values indicate warming). After Sherwood and Nishant (2015) except we use altitude, rather than pressure, for the vertical coordinate. See Methods for further information

Precise quantification of the contemporary abundance of atmospheric CO_2 was initiated by Charles David Keeling at the Mauna Loa Observatory (MLO) in Hawaii during March 1958, as part of the International Geophysical Year (IGY) program (Keeling et al. 1976). On the first day of measurement, an atmospheric CO_2 abundance of 313 parts per million (ppm) was recorded.[12] The MLO CO_2 record is a signature accomplishment of the IGY, carried out from July 1957 to December 1958 and characterized by international cooperation on many scientific fronts. The ability of nations such as the United States and the Soviet Union to collaborate on IGY, despite the Cold War, serves as an inspiration for the level of international cooperation that will be needed to address the consequences of rising GHGs recorded at the Mauna Loa observatory.

On the day this sentence was written, atmospheric CO_2 at MLO was 407.66 ppm. This CO_2 reading amounts to a thirty percent increase relative to Keeling's first

---

[12] Mixing ratio denotes the fraction of all air molecules that exist as a particular compound. Keeling's initial observation means 313 out of every million air molecules were present as CO_2. The history of the Mauna Loa Observatory as well as an account of this initial observation are at https://scripps.ucsd.edu/programs/keelingcurve/2013/04/03/the-history-of-the-keeling-curve

observation and a forty-five percent increase relative to 280 ppm, the atmospheric mixing ratio of CO_2 commonly assumed to have been present at the start of the Anthropocene. Daily measurements of atmospheric CO_2 are provided at various websites, including https://www.co2.earth/daily-co2.

We now describe the scientific evidence that humans are responsible for the rise of CO_2. Our focus is on 1959 to present, the modern instrument era. Readers interested in learning about the human impact on CO_2 over the earlier part of the Anthropocene (i.e., prior to 1959) are encouraged to examine studies such as Siegenthaler and Oeschger (1987), Ruddiman (2003), Le Quéré et al. (2015), and Steffen et al. (2015).

Figure 1.6 shows time series of the atmospheric build-up of CO_2 and fossil fuel emissions of CO_2, from 1959 to present. Measurement of CO_2 at MLO (Keeling et al. 1976) and an estimate of global mean CO_2 provided by the US National Oceanographic and Atmospheric Administration (NOAA) Earth System Research Laboratory (ESRL) (Ballantyne et al. 2012) are shown in Fig. 1.6a. The saw-tooth pattern of the MLO CO_2 reveals the breathing of the biosphere: seasonal minimum occurs in late boreal summer just before deciduous trees, which predominantly exist in the NH, begin to drop their leaves. Seasonal maximum occurs in mid-spring of the NH, just before trees and plants bloom. The global, annual record of CO_2 exhibits a steady upward march over the past six decades.

Figure 1.6b provides our first evidence that humans are responsible for the rise of CO_2 over the past six decades. This panel compares the annual, global release of CO_2 to the atmosphere due to the combustion of fossil fuels (Boden et al. 2013) and land use change (Houghton et al. 2012) (green bars) to the annual rise in global atmospheric CO_2 (blue bars); both quantities are expressed in units of 10^9 metric tons of CO_2 (Gt CO_2) emitted per year.[13] In some years, such as 1977, 1979, 1987, 1988, and 1998, the rise in atmospheric CO_2 is more than half of the CO_2 input to the atmosphere by humans (i.e., the height of the blue bar is more than half the height of the green bar). Typically, the annual rise in the mass of atmospheric CO_2 (blue bars) equals between 40 and 50 % of the mass of CO_2 released to the atmosphere by humans (green bars). This comparison demonstrates *quantitative plausibility* that the observed rise in atmospheric CO_2 during the modern instrument era was indeed due to human activity.

Figure 1.7 illustrates the three most important pieces of observational evidence that scientists use to reveal the *human fingerprint* on rising CO_2. Time series of the mixing ratio of atmospheric CO_2 measured at MLO in Hawaii (19.82°N latitude) are compared to CO_2 measured at the South Pole Observatory (SPO) in Fig. 1.7a. Figure 1.8 compares the difference between annual averages of CO_2 at MLO and SPO ($\Delta CO_2^{MLO-SPO}$) for specific years plotted against the total human release of atmospheric CO_2 for each particular year. Figures 1.7a and 1.8 show that CO_2 is higher in the NH than the Southern Hemisphere (SH). This hemispheric gradient has long been used as evidence for the human influence on atmospheric CO_2, since anthropogenic emissions occur predominantly in the NH (Tans et al. 1990). The strong correlation of $\Delta CO_2^{MLO-SPO}$ versus total human release of CO_2 shown in

[13] CO_2 emissions are usually expressed as either Gt C or Gt CO_2. Here "G" stands for giga, the Greek word for giant, used as an abbreviation for a billion. Emissions in Gt C can be converted to Gt CO_2 by multiplying by 3.664 (Table 1 of Le Quéré et al. (2015)). Here and throughout, we use Gt CO_2 because these units are more convenient for evaluating the Paris Climate Agreement.

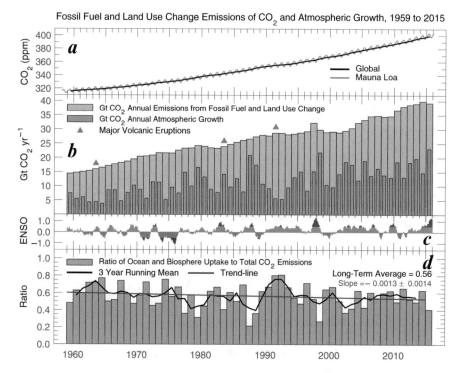

Fossil Fuel and Land Use Change Emissions of CO_2 and Atmospheric Growth, 1959 to 2015

Fig. 1.6 Global carbon cycle, 1959 to present. (**a**) CO_2 mixing ratio from Mauna Loa Observatory (*grey*) (Keeling et al. 1976) and globally averaged CO_2 (*black*) (Ballantyne et al. 2012); (**b**) annual emissions of CO_2 to the atmosphere due to combustion of fossil fuels (Boden et al. 2013) and land use change (Houghton et al. 2012) in units of Gt CO_2 per year (*green bars*), growth of atmospheric CO_2 in the same units (*blue bars*) found from the rise in global annually averaged atmospheric CO_2, and the date of the eruptions of Mount Agung, El Chichón, and Mount Pinatubo (*orange triangles*); (**c**) monthly, Tropical Pacific ENSO Index (Zhang et al. 1997): periods of dark red longer than 5 months indicate an El Niño event; (**d**) ratio of ocean plus biospheric uptake of CO_2 divided by total emissions of CO_2 (i.e., difference between the height of the *green bar* and *blue bars* in panel (**b**) divided by the height of the *green bar*) (*grey*) as well as 3 year running mean (*black*) and trend-line of a linear least squares fit to the 3 year running mean (*blue*). See Methods for further information

Fig. 1.8 further demonstrates that the anthropogenic activity exerts primary control on the hemispheric gradient in atmospheric CO_2 (Fan et al. 1999).

Figure 1.7b illustrates the small decline of atmospheric O_2, which is another fingerprint of the human influence on rising CO_2 (see Methods for an explanation of the units and numbers). As CO_2 is released to the atmosphere by the combustion of fossil fuel, the oxygen (O) content of the newly emitted CO_2 comes from atmospheric molecular oxygen (O_2).[14] If rising atmospheric CO_2 were due to volcanic

---

[14] Fossil fuels are characterized by a mixture of hydrogen (H), carbon (C), and depending on the source other elements such as oxygen (O), nitrogen (N), and sulfur (S). Chemical compositions range from CH_4 (methane or natural gas), C_8H_{18} (octane, commonly used to represent automotive gasoline), to $C_{135}H_{96}O_9NS$ (coal, which can also contain other elements such as arsenic, lead, mercury, etc.). Since H and C are the dominant elements, fossil fuels are commonly called hydrocarbons. The O in CO_2 produced by combustion of methane or gasoline originates *entirely* from

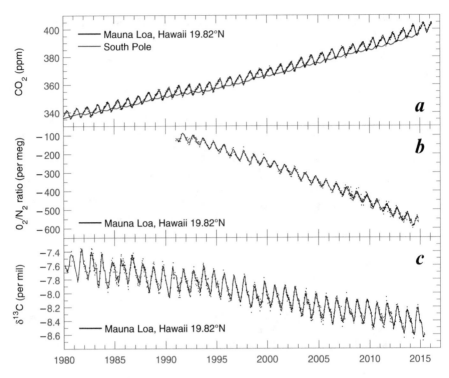

Fig. 1.7 Human fingerprint on rising CO_2. (**a**) Time series of CO_2 mixing ratio from Mauna Loa Observatory (*black*) (Keeling et al. 1976) and the South Pole (Tans et al. 1990); (**b**) ratio of atmospheric O_2 to N_2 measured at the Mauna Loa Observatory in units of per meg, where 1 per meg = 0.00001 % (Keeling et al. 1996); (**c**) abundance of ^{13}C in atmospheric CO_2 at Mauna Loa relative to a standard in units of per mil, where 1 per mil = 0.1 % (Keeling et al. 2005). See Methods for further information

activity or deep sea vents, atmospheric O_2 would be unaffected because the dominant form of outgassed carbon is CO_2, with the O drawn from abundant oxygen in Earth's crust. Measurement of atmospheric O_2 with sufficient precision to quantify the minute, putative decline was an instrumental challenge first overcome by Ralph Keeling (Keeling et al. 1996), the son of Charles David Keeling. The slight decline in atmospheric O_2 recorded in Fig. 1.7b provides strong quantitative evidence that combustion of fossil fuel is the driving factor behind rising CO_2.

The final human fingerprint involves the isotopic composition of atmospheric CO_2. The most common form of carbon has an atomic mass of 12 (^{12}C), due to the presence of six protons and six neutrons in the nucleus.[15] Carbon can exist in two other forms: carbon 13 (^{13}C) and carbon 14 (^{14}C), due to the presence of either 7 (^{13}C) or 8 (^{14}C) neutrons in the nucleus. The chemical properties of a compound are determined by the number of electrons, which equals the number of protons if a compound is neutral

---

atmospheric O_2, whereas the O in CO_2 produced by combustion of coal originates *mainly* from atmospheric O_2.

[15] Atomic mass is the sum of the number of protons and neutrons.

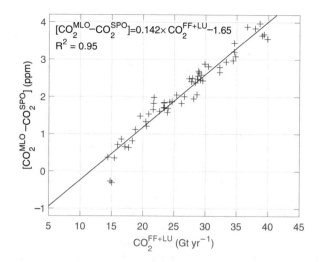

Fig. 1.8 Human fingerprint on hemispheric gradient of CO_2. Difference in annual average CO_2 at Mauna Loa Observatory (MLO) (Keeling et al. 1976) minus CO_2 at the South Pole Observatory (SPO) (Tans et al. 1990) (*vertical axis*) versus annual total emission of CO_2 to the atmosphere from the combustion of fossil fuels (Boden et al. 2013) and land use change (Houghton et al. 2012). Data span the time period 1959–2015. Numerical results of a linear least squares fit as well as the correlation coefficient are also given. See Methods for further information

(i.e., uncharged). Hence ^{12}C, ^{13}C, and ^{14}C are all considered to be different forms of carbon because they all contain six electrons. Most importantly, biological properties of a compound are mass dependent: our bodies prefer ^{12}C over the other two heavier forms of carbon that are digested, because lighter molecules diffuse more readily through our capillaries. The term isotopic composition, as used here, refers to the relative abundance of ^{13}C in a sample of atmospheric CO_2 compared to the sum of ^{12}C, ^{13}C, and ^{14}C in the same sample, and is expressed using the notation $\delta^{13}C$.

Figure 1.7c shows a time series of $\delta^{13}C$ recorded at MLO (Keeling et al. 2005). The downward decline of $\delta^{13}C$ means atmospheric CO_2 is getting isotopically lighter over time. In other words, at the start of the time series in 1980, the relative proportion of ^{13}C to ^{12}C in atmospheric CO_2 at Hawaii was larger than today. This serves as our final fingerprint because the carbon content of fossil fuels, which formed from the decomposition of plants on geologic time-scales, are isotopically light relative to contemporary atmospheric CO_2 (Whiticar 1996). If rising levels of atmospheric CO_2 during the time period shown in Fig. 1.7c had been due primarily to volcanoes, atmospheric CO_2 would have been expected to have gotten isotopically heavier (Rizzo et al. 2014), which is the opposite of what has been observed.

It is stated in Sect. 1.1 that over ***geologic time scales***, atmospheric CO_2 is controlled by volcanic activity and deep sea vents. Yet CO_2 shows no volcanic influence over the time of the modern instrument record. To further illustrate the lack of recent volcanic influence, the orange triangles in Fig. 1.6b have been placed at the time of eruption of Mount Agung, El Chichón, and Mount Pinatubo, the three largest eruptions over the past six decades. The growth of atmospheric CO_2 during the years of these eruptions (1963, 1982, and 1991) is unremarkable compared to other years: in fact, the growth of atmospheric CO_2 (blue bars) after the eruption of Mt. Agung (1963) and Mt. Pinatubo (1991) is suppressed relative to prior years. The best estimate of contemporary release of atmospheric CO_2 by volcanoes and deep sea vents reveals release of about 0.26 Gt of atmospheric CO_2 per year (Marty and Tolstikhin 1998), less than 1 % of the human burden. Interestingly and with touching irony for those who refuse to accept the human influence on global warming, the volcanic release of CO_2 during the 9 h explosive phase of Mt. Pinatubo on 15 June 1991 likely matched the total, global release of CO_2 by humans *on that day* (Gerlach 2011). More than 20,000 days have passed since the start of modern measurements of atmospheric CO_2. On one day and one day only, global human release of atmospheric CO_2 was likely matched by a volcano. Human release of CO_2 has dwarfed volcanic release on the other 19,999 days.

Why have volcanoes been so dominant in the past, yet so unimportant in the present? One factor is that modern human civilization has not yet experienced a volcanic eruption of the magnitude known to have occurred in the past. The Volcanic Explosivity Index (VEI) denotes the size of volcanic eruptions (Newhall and Self 1982), much like the Richter Scale for earthquakes. Mt. Agung, El Chichón, and Mt. Pinatubo had VEIs of 5, 5, and 6, respectively. The most violent eruption Earth has experienced over the past 36 million years was the VEI of 9.1–9.2 eruption of La Garita[16] about 27.8 Mypb (Mason et al. 2004). Since the VEI scale is logarithmic with respect to volume of ejecta, a VEI 9 eruption would eject about 1000 times more mass than Mt. Pinatubo.[17] Had Mt. Pinatubo been VEI 9, it may have matched human emission of CO_2 over the prior 1000 days. In this case, of course, the ejection of CO_2 by such a monstrous event would have been the least of our concerns.

The other factor responsible for the minor role of volcanoes with respect to contemporary atmospheric CO_2 is that Earth is presently in a geologically dormant period. The Deccan Traps of India is one of the largest, most well-studied ancient volcanic features on Earth. Eruptions of this massive province, approximately 65 Mybp, may have been characterized by a decades-long explosive events (Self et al. 2006). The perturbation to atmospheric CO_2 by the Deccan Traps is the subject of active research, with some studies (Dessert et al. 2001) suggesting a considerably larger influence than others (Self et al. 2006).

We conclude by providing a brief overview of the latest understanding of the factors that control atmospheric CO_2. More detailed information, updated annually, is maintained at http://www.globalcarbonproject.org/carbonbudget.

It is well established that a substantial portion of the CO_2 released to the atmosphere by human activity is absorbed by trees and plants (i.e., the terrestrial biosphere) as well as the world's oceans (Le Quéré et al. 2015). Uptake of anthropogenic CO_2 by plants is facilitated by three factors: higher levels of atmospheric CO_2 promote faster growth of plants (Zhu et al. 2016), global warming has increased the length of the growing season (Le Quéré et al. 2015), and human supply of fixed nitrogen to the biosphere promotes a more fertile environment for plant growth (Galloway et al. 2014). The world's oceans

---

[16] The caldera of this ancient volcano is near the town of Creede, Colorado in the United States.

[17] For those with a mathematical background, the calculation is straightforward. The ejected mass is proportional to 10 raised to the power of VEI; therefore, the ratio of mass ejected by a VEI 9 eruption to Mount Pinatubo is 10^9 divided by 10^6, which equals 10^3 or 1000.

contain a mass of carbon about 50 times greater than that in the atmosphere and it has long been known that the world's oceans would uptake a portion of the CO_2 placed into the atmosphere by human activities (Revelle and Suess 1957).

An atmospheric and oceanic phenomena known as El Niño Southern Oscillation (ENSO) has been extensively studied as a driving factor for the variation of the height of the blue bars (atmospheric growth of CO_2) relative to the green bars (human release of CO_2) shown in Fig. 1.6b (Keeling et al. 2005; Randerson et al. 2005; Zeng et al. 2005). When the index shown in Fig. 1.6c is shaded dark red for a period of ~5 months or longer, the tropical ocean/atmosphere system is in the midst of an ENSO event.[18] The growth of atmospheric CO_2 tends to be larger than normal for about a year after the peak of an ENSO event, with the effect maximizing about 6 months after the peak (Zeng et al. 2005). An ENSO event affects atmospheric CO_2 due to suppression of oceanic uptake as well as the tendency for human-set fires to occur in drought stricken regions during certain ENSO years (Randerson et al. 2005). During late 2015, Earth experienced another major ENSO event, which likely was responsible for the more rapid rise of atmospheric CO_2 in 2015 compared to prior years. Indeed, the preliminary estimate of total human release of CO_2 in year 2015 given by Le Quéré et al. (2015), which is the origin of the last green bar in Fig. 1.6b, shows a slight decline relative to 2014. Should this decline in human release of CO_2 continue in future years, the height of the blue bars in future updates to Fig. 1.6b will fall relative to the value for 2015, except for years marked by either large ENSO events and/or extensive biomass burning.

The fraction of anthropogenic CO_2 removed each year via the world's terrestrial biosphere and oceans is depicted by the grey bars in Fig. 1.6d. There is considerable year-to-year variability, which has been widely studied and is attributed mainly to terrestrial biosphere (Bousquet et al. 2000; Le Quéré et al. 2003). Averaged over the entire data record, 56 % of the CO_2 released to the atmosphere by humans by the combustion of fossil fuels and land use change has been absorbed by land and ocean sinks. In other words, the actual rise in atmospheric CO_2 equals about 44 % of that known to have been emitted by humans.

The efficiency of the combined land and ocean sink for atmospheric CO_2 appears to be weakening over time. Figure 1.6d contains two lines. One shows a 3 year running mean (black) of the numerical values of each grey bar, for data starting in 1959 and ending in 2014. Values for 2015 are excluded from the 3 year running mean, because data for this year are considered preliminary at the time of writing. An entity such as a 3 year running mean is a common statistical method used to analyze data that exhibit a large amount of year-to-year variability, such as the grey bars in Fig. 1.6d. The trend-line (blue) shows a linear least squares fit to the 3 year running mean, another common technique used to examine geophysical data. The trend-line has a slope of -0.0013 per year, which means the fraction of anthropogenic CO_2 removed by the combined land and ocean sink may have declined from about 0.6 in 1959 to about 0.53 in 2014. However, there is considerable uncertainty (in this case,

---

[18] During an ENSO event warm waters in the Tropical Western Pacific ocean migrate to the Central and Eastern Pacific, causing shifts in the location of oceanic upwelling and atmospheric storms, as well as significant perturbations to the global carbon cycle. An informative animation of ENSO is provided at http://esminfo.prenhall.com/science/geoanimations/animations/26_NinoNina.html

±0.0014 per year) in the slope of the trend fit line. This uncertainty encompasses (albeit, just barely) the possibility that the combined land and ocean sink may not have actually changed and also, at the same time, another possibility that this uptake could have changed from 0.6 in 1959 to 0.45 in 2014. This analysis builds upon the work and supports the conclusions of Le Quéré (2010), who also emphasize the urgent need to reduce the uncertainty in the time rate of change of the combined land and ocean sink for human release of atmospheric CO_2. If the efficiency of the combined land and ocean sink for CO_2 is truly declining over time, then this is enormously important for the response of society to anthropogenic release of GHGs.

1.2.3.3   Methane

Methane (CH_4) is a vitally important anthropogenic GHG. The atmospheric abundance of CH_4 has risen from a pre-Anthropocene value of 0.7 ppm to a contemporary abundance of 1.84 ppm (Fig. 1.2). The rise in CH_4 between 1750 and 2011 has induced a RF of climate of 0.48 W m^{-2} (Fig. 1.4), second only to the RF of CO_2 among anthropogenic GHGs.[19] Methane is therefore commonly referred to as the second most important anthropogenic GHG.

Studies of atmospheric CH_4 are numerous, complex, and quite varied, owing to a variety of natural and human sources (see Kirschke et al. (2013) and references therein). Figure 1.9 shows an estimate of the sources (i.e., flux into the atmosphere) of CH_4, in units of 10^{12} g of CH_4 (Tg CH_4) emitted per year,[20] averaged over the decade 2000–2009 from Conrad (2009) and Kirschke et al. (2013). The figure also contains an estimate of the sinks (i.e., atmospheric loss) of CH_4 over the same period of time.

A number of scientifically important details regarding atmospheric CH_4 are contained in Fig. 1.9. First, the magnitude of the source is slightly larger than the sink, consistent with the fact that atmospheric CH_4 is rising. Also, there are various human and natural sources of considerable magnitude. As noted above, wetlands are the largest natural source of CH_4. Other natural sources include termites and the release of CH_4 from gas hydrates.[21] Finally, anthropogenic production of CH_4 occurs due to many aspects of our industrialized world, including the fossil fuel industry,

---

[19] The RF of climate due to CO_2 over the same time period was 1.82 W m^{-2}. The notion that CH_4 is a more potent GHG than CO_2 is reconciled with these two RF estimates upon realization that the rise of the atmospheric mixing ratio of CO_2 over the Anthropocene, 120 ppm, is about 106 times the rise of CH_4. For those who would like to dig into the numbers, radiative efficiencies of CO_2 and CH_4 are needed. In mixing ratio units, these radiative efficiencies are 1.4×10^{-2} W m^{-2} per ppm for CO_2 and 3.7×10^{-1} W m^{-2} per ppm for CH_4 (see Table TS.2 of IPCC (2007)). A "back of the envelope" estimate for the expected RF due to CH_4 is then:

$$[1.82 \text{ W m}^{-2} \times (3.7 \times 10^{-1} \div 1.4 \times 10^{-2})] \div 106 = 0.45 \text{ W m}^{-2}.$$

This estimate for the RF of CH_4 over the Anthropocene is quite close to the actual IPCC (2013) value of 0.48 W m^{-2}, which was found in a much more computationally intensive manner.

[20] Tera is derived from the Greek word teras, meaning monster, and is often used as a prefix to denote 10^{12}, or a trillion. A mass of 1 Tg (10^{12} g) is the same as one thousandth of a giga tonne, where tonne refers to metric ton.

[21] Methane hydrates are water ice structures that contains gaseous CH_4 in the core, and are prevalent in continental margins (Kvenvolden 1993).

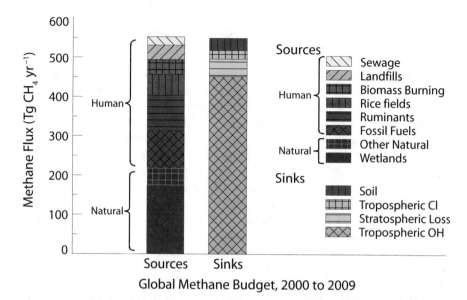

Fig. 1.9 Global methane budget. Source and sinks of atmospheric methane, over the decade 2000–2009, expressed as flux either into or out of the atmosphere. After Conrad (2009) and Kirschke et al. (2013). Human and natural sources, as well as components of all terms, are indicated. See Methods for further information

human set fires (biomass burning), microbial processes in the stomachs of ruminants,[22] as well as anaerobic conditions common in rice paddies and landfills. As detailed in Chap. 4, if human release of methane is to be curtailed, many aspects of modern society will need to be addressed, including how we heat our homes, generate our electricity, and produce our food.

Two considerable complications for the proper accounting of the human release of CH_4 are posed by possible alteration of the wetland source due to climate-change induced changes of the hydrologic cycle (i.e., floods and drought) as well as the possible release of prodigious amounts of CH_4 from the Arctic as permafrost thaws, due to global warming (Koven et al. 2011). For now, at least, the source of CH_4 due to Arctic permafrost is small on a global scale (Kirschke et al. 2013).

We turn our attention to the scientific importance of the numerical estimates of the CH_4 source and sink strengths shown in Fig. 1.9. The globally averaged sink for CH_4 is 550 Tg per year. The mass of CH_4 in the atmosphere, at present,[23] is about 5326 Tg. The atmospheric lifetime of CH_4 is given by:

---

[22] Ruminants are mammals such as cattle, sheep, deer, giraffes, etc. that acquire nutrients by fermenting plant-based foods in a specialized stomach prior to digestion.

[23] We can approximate the mass of CH_4 in the atmosphere by multiplying the mass of the entire atmosphere, 5.2×10^{21} g, by the mixing ratio of CH_4, which is 1.84 ppm or 1.84 out of every million air molecules. We must also account for the ratio of the atomic mass of CH_4 (16) to the mean atomic mass of air (28.8). The atmospheric mass of CH_4 therefore equals $1.84 \times 10^{-6} \times 5.2 \times 10^{21}$ g $\times (16/28.8) = 5.326 \times 10^{15}$ g $= 5326$ Tg.

$$\tau_{CH4} = \frac{Mass}{Removal\ Rate} = \frac{5326\mathrm{Tg}}{550\mathrm{Tg\ year^{-1}}} = 9.7\,\mathrm{year} \qquad (1.7)$$

On average, a molecule of CH_4 released to the atmosphere will persist for about a decade until it is removed by either a chemical reaction or a soil microbe. Prather et al. (2012) report a present-day CH_4 lifetime of 9.1 ± 0.9 years, consistent with our calculation above. Conversely, a lifetime of 12.4 years for CH_4 was used by IPCC (2013) in the calculation of GWPs because it is thought that the release of a large pulse of CH_4 to the atmosphere will prolong the atmospheric lifetime, due to resulting changes in the chemical composition of the tropical troposphere.

The ~10 year atmospheric lifetime for CH_4 has important policy implications. This is best illustrated by comparing the human release of CH_4 to that of CO_2. Throughout the world, humans presently release about 335 Tg of CH_4 and 39 Gt of CO_2 per year. Since 1000 Tg = 1 Gt, these sources are 0.335 Gt of CH_4 and 39 Gt of CO_2 per year: i.e., the mass of CO_2 released to the atmosphere each year by human society is about 116 times more than the mass of CH_4. The impact on climate is entirely dependent on the time scale of interest. Nearly all of the CH_4 released to the atmosphere in year 2015 will be gone by the end of this century. The **CO_2-equivalent** emission of CH_4, found by multiplying the current release by the GWP for CH_4 for a 100-year time horizon, is 28×0.335 Gt of CH_4 or 9.4 Gt per year. If our concern is global warming over the next century, then we would conclude the human release of CO_2 in year 2015 was about four times more harmful for climate ($39 \div 9.4 = 4.1$) than the release of CH_4. However, if our concern is the next two decades, we must consider the GWP of CH_4 over a 20-year time horizon. In this case, the CO_2-equivalent emission of CH_4 is 84×0.335 Gt or 28.1 Gt per year, and we would conclude the present human release of CH_4 is nearly as harmful for climate (28.1 versus 39) as the release of CO_2.

As noted above, international policy for the regulation of GHGs generally utilizes GWPs found over a 100-year time horizon. Perhaps this is appropriate, given CO_2 is such a long-lived GHG (i.e., a CO_2 molecule released today by humans will likely persist in the atmosphere longer than a molecule CH_4). However, should the world ever face an impending climate catastrophe in the midst of rapidly rising abundances of both atmospheric CO_2 and CH_4, the greatest leverage for near-immediate relief will be to reduce anthropogenic emissions of CH_4 (Shindell et al. 2009) or other short-lived pollutants (Pierrehumbert 2014). Of course this is much easier stated than accomplished given the wide variety of human activities that release CH_4, as well as the tendency for energy production in the United States to become increasingly more CH_4-based, given the abundant source of natural gas now being extracted by fracking.[24]

---

[24] The extraction of CH_4 by the hydraulic fracturing ("fracking") of ancient shale following horizontal drilling has led to a recent, major rise in production of this fossil fuel. This is discussed further in Chap. 4.

Considerable research effort has been directed towards quantification of the anthropogenic versus natural sources of atmospheric CH_4 due to the scientific importance of this apportionment. Suppose, as is likely, that CH_4 had the same (or nearly the same) lifetime for removal from the atmosphere as today, for conditions that prevailed prior to the Anthropocene. Also, if the natural source of atmospheric CH_4 was the same (or similar) for these two time periods, it can be shown that:

$$\frac{CH_4^{\text{Pre-Anthropocene}}}{CH_4^{\text{Present}}} = \frac{Source^{\text{Natural}}}{Source^{\text{Natural+Human}}} \qquad (1.8)$$

where $Source^{\text{Natural}}$ is the present natural flux of CH_4 to the atmosphere and $Source^{\text{Natural+Human}}$ is the total flux. Using the numerical values for these two fluxes[25] from Fig. 1.9, which are based on Table 1 of Kirschke et al. (2013), yields an estimate for $CH_4^{\text{Pre-Anthropocene}}/CH_4^{\text{Present}}$ of 0.39. This estimate is astonishingly close to the actual ratio of $CH_4^{\text{Pre-Anthropocene}}/CH_4^{\text{Present}} = 0.38$, found using the atmospheric abundances given in the opening paragraph of this section.[26] Thus, an analysis of the sources of atmospheric CH_4 for the contemporary atmosphere provides *strong quantitative support* for the notion that human activities are indeed responsible for the rise of atmospheric CH_4 over the course of the Anthropocene.

1.2.3.4 Nitrous Oxide, Ozone, and Ozone Depleting Substances

Nitrous oxide (N_2O) is commonly considered to be the third most important anthropogenic GHG. The atmospheric abundance of N_2O has risen from a pre-Anthropocene value of 0.273 ppm to a contemporary abundance of 0.329 ppm (Fig. 1.2). The rise in N_2O between 1750 and 2011 has induced a RF of climate of 0.17 W m^{-2} (Fig. 1.4).

Nitrous oxide is long-lived, with a lifetime of about 120 years. The vast majority of the atmospheric loss of N_2O occurs in the stratosphere (Minschwaner et al. 1993). Nitrous oxide has a GWP of 264 (20-year time horizon) or 265 (100-year time horizon) (Table 1.1) according to IPCC (2013). The GWP of N_2O is nearly the same for both time horizons because a pulse of N_2O released to the atmosphere decays, within models used to calculate GWPs, in a manner quite similar to the decay of a pulse of CO_2.

Current best understanding of the human sources of N_2O is described in Chap. 6 of IPCC (2013). The total anthropogenic source is estimated to be 21.7 Tg of N_2O per year,[27] albeit with considerable uncertainty. The human source could be as low

---

[25] $Source^{\text{Natural}} = 218$ Tg year^{-1} (total of the human terms; i.e., height of the six rectangles to the right of "Human" in Fig 1.9); $Source^{\text{Natural+Human}} = 553$ Tg year^{-1} (total of all sources, Fig. 1.9).

[26] These abundances yield $CH_4^{\text{Pre-Anthropocene}}/CH_4^{\text{Present}} = 0.7$ ppm/1.84 ppm = 0.38.

[27] The IPCC (2013) best estimate for human release of N_2O is 6.9 Tg of nitrogen per year, but we must convert to N_2O to make use of the GWP of N_2O. 6.9 Tg of N per year is the same as $6.9 \times (44 \div 14) = 21.7$ Tg of N_2O per year, where 44 and 14 are the atomic masses of N_2O and N.

as 8.5 or as high as 34.9 Tg N_2O per year according to Table 6.9 of IPCC (2013). Prather et al. (2012) report a smaller best estimate for the human source of 20.4 Tg N_2O per year, with a reduced uncertainty of ±4 Tg N_2O per year. Agriculture is the dominant activity responsible for human release of N_2O: use of nitrogen fertilizers results in release of N_2O to the atmosphere due to microbial processes in soils (Smith et al. 1997). Contemporary human emissions of N_2O presently make a contribution to global warming[28] that is ~15 % that of emissions of CO_2.

The largest natural sources of N_2O are production from soils that lie beneath vegetation unperturbed by humans and release from the world's oceans. The natural source is estimated to be 34.6 Tg of N_2O per year, again with considerable uncertainty (range from 17.0 to 61.6 Tg of N_2O per year) (IPCC 2013). Large uncertainties for both the human and natural sources of N_2O, as well as the long atmospheric lifetime for N_2O, preclude meaningful use of Eq. 1.8 to examine the consistency between the rise in N_2O and our understanding of the natural and anthropogenic source strengths. Nonetheless, the long-term rise in N_2O since 1977, the observation of larger abundances in the NH than the SH documented on websites such as http://www.esrl.noaa.gov/gmd/hats/combined/N2O, and field measurements of strong anthropogenic sources (Table 6.9 of IPCC (2013)) all provide strong scientific evidence that humans are responsible for the vast majority of the rise in N_2O over the course of the Anthropocene.

The possible increase in atmospheric N_2O due to expanded use of biofuels will receive considerable attention in the next few decades. There is considerable interest in the development of biofuels as a replacement for fossil fuels because, in theory, biofuels could be close to carbon neutral. The notion of carbon neutrality is predicated on the fact that the carbon in a hydrocarbon fuel produced by recent photosynthesis has been drawn out of the atmosphere just prior to combustion: i.e., the carbon is recycled. One of the many concerns regarding the modern biofuel industry is that the associated increase in production of atmospheric N_2O due to the need for additional fertilizer will offset the climate benefit from the supposed carbon neutrality of this new fuel source (Crutzen et al. 2016).

The effect of N_2O on stratospheric O_3 will also likely receive attention by researchers. Loss of N_2O occurs in the stratosphere and, upon decomposition, N_2O produces compounds that deplete stratospheric ozone (Ravishankara et al. 2009). Most interestingly, the ozone depletion potential[29] of N_2O depends on future atmospheric abundances of CO_2 and CH_4 (Revell et al. 2015). Not only are CO_2, CH_4, and N_2O (as well as chlorofluorocarbons, or CFCs) all important for climate, but these compounds are also inextricably linked for the future recovery of Earth's ozone layer.

---

[28] Recalling that 1000 Tg = 1 Gt, the human release of CO_2 is 39 Gt C per year, and making use of a GWP for N_2O of 264 results in the following calculation for the contribution of N_2O to global warming relative to that of CO_2: $[21.7 \text{ Tg year}^{-1} \div 1000 \text{ Tg/Gt}] \div [39 \text{ Gt year}^{-1}] \times 264 = 0.15$.

[29] Ozone depletion potential is a metric developed by atmospheric chemists to gauge the harmful effects of various compounds on the stratospheric ozone layer.

While it is certainly true that most scientists consider N_2O to be the third most important anthropogenic GHG, it is worth noting that the contribution to the RF of climate over the course of the Anthropocene by N_2O is smaller than that of both Ozone Depleting Substances (ODS) and tropospheric O_3 (Fig. 1.4). Why then is N_2O commonly considered to be the third most important anthropogenic GHG? The answer is nuanced but provides insight into the multi-disciplinary nature of modern atmospheric science.

The category labeled ODS in Fig. 1.4 consists of many gases (see Methods), including numerous CFCs, hydrochlorofluorocarbons (HCFCs), carbon tetrachloride (CCl_4), methyl chloroform (CH_3CCl_3), etc. Industrial production of this class of compounds has been successfully regulated by the Montreal Protocol and subsequent amendments due to the harmful effects of these chemicals on Earth's protective ozone layer (WMO 2014). It is not commonly appreciated, but the climate protection accomplished by the Montreal Protocol (due to reduction in the atmospheric abundance of ODS that would have otherwise occurred) far exceeds the climate protection accomplished by the Kyoto Protocol (Velders et al. 2007). In other words, the positive RF of climate due to ODS in Fig. 1.4 would have been much larger had industrial production of these compounds not been halted by the Montreal Protocol. Nonetheless, most scientists do not apply the GHG label to the class of chemical compounds that deplete Earth's ozone layer. Also, none of the ODS compounds, alone, has a RF of climate as large as N_2O. So N_2O survives this challenge to its third place status.

The category labeled tropospheric O_3 in Fig. 1.4 also exerts a RF of climate that exceeds that due to N_2O. Over the course of the Anthropocene, human release of chemicals such as carbon monoxide and nitrogen oxides produced by biomass burning and the combustion of fossil fuels has led to a build-up of tropospheric ozone, exerting a considerable influence on the RF of climate (Fig. 1.4). There has also been a slight cooling effect to the decline in stratospheric O_3 over the Anthropocene. Lack of consideration of tropospheric O_3 as the third most important anthropogenic GHG is due to various factors, including: (a) tropospheric O_3 is not emitted directly by humans but rather is produced in the atmosphere following chemical reactions of O_3 precursors released by humans; (b) surface O_3, which is an important subcategory of tropospheric O_3, is regulated by air quality agencies throughout the world (i.e., O_3 poses more harm to air quality than to climate); (c) all of the other anthropogenic GHGs tend to be long lived (atmospheric lifetimes greater than a year) and have nearly uniform global distributions, whereas tropospheric O_3 is short lived (atmospheric lifetime of minutes to hours) and is highly variable. In the minds of most climate scientists, N_2O survives the challenge from tropospheric O_3 to its third place ranking among anthropogenic GHGs.

We conclude this section by noting the radiative forcing of climate due to tropospheric ozone is due mainly to enhancements over background levels in the tropical upper troposphere (Shindell and Faluvegi 2009). Elevated levels of ozone in this region of the atmosphere are mainly due to biomass burning (Anderson et al. 2016). It is therefore likely that air quality regulations in the developed world will have little effect on the RF of climate due to tropospheric O_3, since so much of the developed world is outside of the tropics. Reducing human set fires in the tropics is a vexing problem, given that many of the fires are set to clear land for agriculture. Nonetheless, the development of effective controls on human set fires will likely be necessary to reduce the RF of climate due to tropospheric O_3 (Keywood et al. 2013).

1.2.3.5   HFCs, PFCs, SF_6 and NF_3

Sulfur hexafluoride (SF_6) and the class of compounds called hydrofluorocarbons (HFCs) and perfluorocarbons (PFCs) often appear in the climate regulation lexicon because these compounds, along with CO_2, CH_4, and N_2O, were all considered by the original Kyoto Protocol. Over the course of the Anthropocene, it is estimated that the RF of climate due to SF_6, HFCs, and PFCs has been about 0.03 W m^{-2} (Other F-gases, Fig. 1.4), which is about 1 % of the total RF of climate due to all anthropogenic GHGs. Nonetheless, there is concern the RF of climate of these compounds could rise in the future (IPCC/TEAP 2005; Velders et al. 2009; Zhang et al. 2011). The Doha amendment, adopted in December 2012, added nitrogen trifluoride (NF_3) to the list of GHGs in the Kyoto Protocol. As such, we'll provide a brief description of the lifetimes, GWP, and industrial uses of HFCs, PFCs, SF_6, and NF_3.

First a little demystification of the chemistry. All of the compounds considered in this section contain at least one fluorine (F) atom, which is in the halogen column of the periodic table. Also and most importantly, none of the compounds discussed here contain any chlorine or bromine atoms. Chlorine (Cl) and bromine (Br), two other halogens, are harmful to Earth's ozone layer and any industrial compound containing either Cl or Br that has a long enough lifetime to reach the stratosphere falls under the auspices of the Montreal Protocol. Natural production of HFCs, PFCs, SF_6 and NF_3 does not occur. Therefore, the presence of these compounds in the atmosphere at a detectable level is attributed to human activity.

The HFCs, PFCs, SF_6 and NF_3 group of GHGs are chemically stable and radiatively active. Most of these compounds have either a single central element surrounded by either numerous fluorine atoms or some combination of fluorine and hydrogen atoms, or a central double carbon similarly surrounded. These chemicals have various physical properties that have resulted in a wide range of industrial applications. The molecular structure of these compounds makes them very long lived: most survive intact until they encounter the intense ultraviolet radiation environment of Earth's upper stratosphere, except for some HFCs that are removed by chemical reactions in Earth's troposphere. Finally, the presence of F in these molecules creates what scientists call a strong dipole moment. These dipole moments tend to occur at wavelengths where thermal radiation emitted by Earth's surface would otherwise escape to space (i.e., an atmospheric window). Chemicals that are long-lived and absorb in an atmospheric window tend to have large GWPs. Typically, the more F in a compound, the higher the GWP (Bera et al. 2009).

Table 1.2 gives the GWPs (100-year time horizon), atmospheric lifetimes, and industrial uses of HFCs, PFCs, SF_6, and NF_3. The information is based on Table 8.A.1 of IPCC (2013) and is intended to serve as a synopsis of this longer table, which

Table 1.2 Properties of long lived HFCs, PFCs, SF_6, and NF_3

GHG	GWP[a]	Lifetime (years)	Industrial use
HFCs	116–12,400	1.3–242	Refrigeration, foam blowing, and by product of manufacturing of HCFCs
PFCs	6290–11,100	2000–50,000	Aluminum smelting
			Semi-conductor manufacturing
SF_6	23,500	3200	Insulator in high voltage electrical equipment
			Magnesium casting
			Semi-conductor manufacturing
NF_3	16,100	500	Semi-conductor manufacturing

[a]For 100-year time horizon

spans four pages, covers more than 100 compounds, and contains many properties for each compound. Only ranges of GWPs and lifetimes are given for HFCs and PFCs in Table 1.2. The atmospheric lifetime for some of these molecules is remarkably long (CF_4, a PFC, has a lifetime of 50,000 years) and many of the GWPs are huge (C_2F_6, another PFC, has a GWP of 11,100).

Hydrofluorocarbons (HFCs) reside at the intersection of ozone depletion and global warming. The Montreal Protocol, which was enacted to protect Earth's ozone layer, guided a transition from industrial production of CFCs to a class of gases called hydrochlorofluorocarbons (HCFCs), because HCFCs are less harmful to the O_3 layer than CFCs.[30] The Montreal Protocol requires a further transition from HCFCs to hydrofluorcarbons (HFCs) because, as noted above, HFCs pose no threat to the ozone layer. However, the GWPs of HFCs (Table 1.2) generally far exceed the GWPs of HCFCs (Table 8.A.1, IPCC (2013).

The future RF of climate due to HFCs is uncertain. Velders et al. (2009) project the RF of climate due to HFCs could be 0.4 W m^{-2} by mid-century, considerably larger than the RF due to HFCs considered by IPCC (2013). The primary reason for this difference is their projection of considerably larger growth in the atmospheric abundance of HFC-125 (formula CHF_2CF_3; lifetime = 28 years; GWP = 3170) than in the scenarios used to guide the IPCC climate models.

A number of scientists and policy-makers have lobbied for HFCs to be removed from the UNFCCC basket of GHGs and placed under the auspices of the Montreal Protocol. The argument for this transition is twofold: (1) the production of HFCs was initiated by the Montreal Protocol; (2) this governing body has been extraordinarily effective due to close cooperation between atmospheric scientists, the chemical manufacturing industry, and policy members who staff the Parties of the

---

[30] CFCs are a class of chemicals that contain chlorine, fluorine, and carbon atoms, whereas HCFCs are a class of chemicals that contain hydrogen, chlorine, fluorine, and carbon atoms. In some ways, bookkeeping would be easier had the former been labeled ClFCs and the latter HClFCs. Alas, the first "C" in these compounds stands for chlorine and the second stands for carbon. To make matters more confusing, HFCs are chemicals that contain only hydrogen, fluorine, and carbon atoms. Here the "C" stands for carbon. So if the C comes after the F, it stands for carbon.

Montreal Protocol. Also, it is worth noting that it is inconceivable that the gross domestic product of any country could be adversely affected by regulation of HFCs. In other words, the stakes for the world's economies are low with regard to regulation of HFCs. On 15 October 2016, at the 28th Meeting of the Parties of the Montreal Protocol held in Kigali, Rwanda, an agreement was reached to regulate the future production of HFCs under the Montreal Protocol. This marks the first time the Montreal Protocol has had direct authority over a class of chemical compounds that pose no threat to the ozone layer.

Perfluorocarbons (PFCs) are a class of compounds containing only carbon and fluorine that resist heat, oils, and staining. The most abundant PFCs are PFC-14 (CF_4), PFC-116 (C_2F_6), and PFC-218 (C_3F_8). Atmospheric levels of these compounds have risen steadily; contemporary levels of CF_4, C_2F_6, and C_3F_8 are a factor of 2, 4, and 10 larger, respectively, than observed during the onset of observations in the early 1970s (Mühle et al. 2010). It has been projected that the RF of climate due to all PFCs could approach 0.04 W m^{-2} by end of this century (IPCC/TEAP 2005; Zhang et al. 2011). While this would represent only a small contribution to global warming, PFCs will continue to be monitored due to their extremely long atmospheric lifetimes (Table 1.2).

Sulfur hexafluoride (SF_6) is an excellent insulator favored in the high voltage, electric industry because this compound is non-flammable.[31] The atmospheric abundance of SF_6 has risen steadily since the early 1970s and shows no sign of abating (Rigby et al. 2010). It has been estimated that the RF of climate due to SF_6 could reach 0.037 W m^{-2} by the end of the century (Zhang et al. 2011). As for PFCs, SF_6 bears monitoring due to its atmospheric lifetime of 3200 years (Table 1.2).

The sulfur and fluorine compound sulfuryl fluoride (SO_2F_2) is used as an insecticide and is also monitored, due to a GWP of 800. However, SO_2F_2 has a lifetime of only 36 years. As a result, atmospheric abundances would decline relatively soon after any corrective action were taken, if such action were ever needed.

The sulfur, fluorine, and carbon containing compound SF_5CF_3 received considerable attention in the media following discovery of a surprisingly large atmospheric abundance (Sturges et al. 2000). This gas was termed a "super GHG" because it has the highest radiative efficiency, 0.57 W m^{-2} ppb^{-1}, of any GHG ever studied. However, recent measurements reveal a slowdown in the emissions to the atmosphere (Sturges et al. 2012) and the present RF of climate of SF_5CF_3 is a miniscule ~8.6×10^{-5} W m^{-2}.[32]

Finally, nitrogen trifluoride (NF_3) is the latest member of the GHG-club. In 2008, several studies appeared calling attention to the RF of climate due to this previously unappreciated compound (Prather and Hsu 2008; Tsai 2008). As noted above, NF_3 was added to the Kyoto Protocol list of GHGs as part of the Doha amendment in 2012. The lifetime and GWP of NF_3 are given in Table 1.2. The primary atmo-

---

[31] At one time SF_6 was used to cushion sports shoes, but this use ceased a decade ago and is not considered to be atmospherically important.

[32] Atmospheric abundance of SF_5CF_3 was 0.00015 ppb in 2012 (Sturges et al. 2012); RF of SF_5CF_3 =0.00015 ppb×0.57 W m^{-2} ppb^{-1}=8.6×10^{-5} W m^{-2}.

spheric release of NF_3 seems to be due to the manufacture of large, liquid crystal display screens (Thomas et al. 2012). The present RF of climate due to NF_3 is small, $\sim 2.4 \times 10^{-4}$ W m^{-2}.[33] Perhaps this late-comer to the GHG-club will one day be known as the couch potato GHG.

1.2.3.6   Aerosols

Aerosols are small solid or liquid particles suspended in air. In the context of this book, we use aerosols to refer to particles either emitted directly into the atmosphere by a particular human activity (typically fossil fuel combustion or fires) or particles that form following chemical and physical transformations in the atmosphere of pollutants known as aerosol precursors. The only natural aerosols we shall consider are those resulting from volcanic eruptions; volcanic aerosols only affect climate if they exist in the stratosphere.

Aerosols, particularly those containing the element sulfur, reflect incoming solar radiation, which cools the surface. Sulfate aerosols tend to be produced from pollutants emitted by coal-fired power plants, ships, and diesel fueled trucks and cars, although there is a strong movement towards use of ultra-low sulfur diesel fuel in some parts of the world (Krotkov et al. 2016). Volcanic aerosols, which exert short-term climatic cooling, are also composed of sulfate (Lacis and Mischenko 1995). Sooty aerosols, termed black carbon, are likewise produced by combustion of fossil fuels and biomass burning. Black carbon aerosols have a warming effect because these particles absorb solar radiation (Bond et al. 2013).

The association of human activity with the presence of tropospheric aerosols is well established from both ground-based (Jimenez et al. 2009; Yoon et al. 2016) and space-based observations (Streets et al. 2013; Yoon et al. 2014; He et al. 2016; McLinden et al. 2016). Yet, quantification of the RF of climate due to tropospheric aerosols continues to pose a scientific challenge due to the inability to precisely define numerical values of both the direct modulation of RF by anthropogenic aerosols (Myhre 2009; Kahn 2012; Bond et al. 2013) and the changes in RF driven by the effect of aerosols on clouds (Morgan et al. 2006; Carslaw et al. 2013). The IPCC (2013) best estimate and uncertainty of ΔRF over the course of the Anthropocene for these two terms, labeled Aerosol Direct Effect and Aerosol-Cloud Interaction, are shown in Fig. 1.4.

Tropospheric aerosols lie at the nexus of public health, air quality, and climate change. Exposure to small (Dominici et al. 2006) and/or toxic (Bell et al. 2007) aerosols has deleterious effects on human health. As a consequence, movements are underway throughout the world to reduce both the direct emission of aerosols as well as the emission of aerosol precursors. Reductions in the abundance of tropospheric aerosols and aerosol precursors, in response to air quality legislation motivated by public health concerns, have been readily observed by space-borne

---

[33] Atmospheric abundance of NF_3 peaked at 0.0012 ppb in late 2011 (Arnold et al. 2012) and radiative efficiency is 0.2 W m^{-2} ppb^{-1} (Table 8.A.1 of IPCC (2013); RF of NF_3=0.0012 ppb×0.2 W m^{-2} ppb^{-1} =2.4×10^{-4} W m^{-2}.

instrumentation throughout the world (Streets et al. 2013; Yoon et al. 2014; He et al. 2016). As such, the climate system is presently transitioning from an era where the cooling of climate due to aerosols may have had close to comparable strength as GHG induced warming to an era where the radiative warming due to GHGs will dominate aerosol cooling (Smith and Bond 2014).

The transition to a GHG dominated regime is illustrated in Fig. 1.10. This figure shows ΔRF due to CO_2, CH_4, and N_2O as well as all anthropogenic GHGs from 1850 to 2100 for the Representative Concentration Pathway (RCP) 4.5 scenario (Thomson et al. 2011) used throughout IPCC (2013). Total ΔRF due to all anthropogenic GHGs reaches 4.5 W m^{-2} in year 2100, as designed. The error bars for the ΔRF terms of GHGs, placed at year 2011, are from IPCC (2013). These uncertainties represent 5 and 95 % confidence intervals.

Figure 1.10 shows 71 plausible values for time series of ΔRF due to tropospheric aerosols published by Smith and Bond (2014). The colors correspond to least cooling (reds) to most cooling (blues); the black line denotes the central (median) scenario. These estimates are based on time series of the direct RF of climate due to black carbon, organic carbon, and sulfate aerosols as well as the effect of aerosols on clouds, all tied to the emissions of aerosols and aerosol precursors from the RCP 4.5 scenario. There exists considerable uncertainty with each of these terms. Most importantly, these uncertainties are handled in a self-consistent manner for each of the 71 scenarios over the time period 1850–2100. The scenarios colored in red (least cooling) assume black carbon aerosols exert considerable warming of climate, offsetting nearly all of the cooling by sulfate and organic carbon and the effect of aerosols of clouds. Conversely, the scenarios colored in blue (most cooling) assume black carbon aerosols exert little warming and that sulfate plus organic carbon, combined with the cloud response have led to about 1.4 W m^{-2} cooling in year 2011. For these large cooling scenarios, tropospheric aerosols offset nearly half of the ~2.8 W m^{-2} warming due to GHGs in year 2011.

The difference between the blue and red curves represents the uncertainty in the radiative forcing of climate due to aerosols. As we shall see in Chap. 2, this uncertainty limits our ability to forecast future global warming. All of the aerosol scenarios converge to near zero ΔRF in year 2100. Forecast values of ΔT in 2100 depend on ΔRF from GHGs (known well, provided CO_2, CH_4, N_2O, and the minor GHGs are specified) combined with the true value of climate feedback (see Sect. 2.2.1.2). The climate record over 1850 to present can be fit nearly equally well under two contrasting scenarios: (i) the true value of aerosol RF happened to be little cooling (red curves, aerosols, Fig. 1.10) in which case climate feedback must be modest; (ii) the true value of aerosol RF happened to be large cooling (blue curves) in which case climate feedback must be considerable. If we assume the feedback inferred from the climate record persists over time, then the future rise in ΔT for the modest feedback scenario will be considerably smaller than the future rise in ΔT for the considerable feedback scenario. Even though the human fingerprint on tropospheric aerosol loading is extremely well established, uncertainty in the climatically critical quantity ΔRF due to aerosols leads to considerable spread in future projections of global warming.

Fig. 1.10 The rise and fall of RF due to aerosols. Time series of radiative forcing of climate (ΔRF) due to CO_2, CH_4, N_2O, and all anthropogenic GHGs, from 1850 to 2100, based on the RCP 4.5 scenario (Meinshausen et al. 2011) (*top*) and 71 plausible scenarios for total ΔRF due to anthropogenic aerosols (combination of the aerosol direct effect and the aerosol-cloud interaction) from Smith and Bond (2014) (*bottom*). See Methods for further information

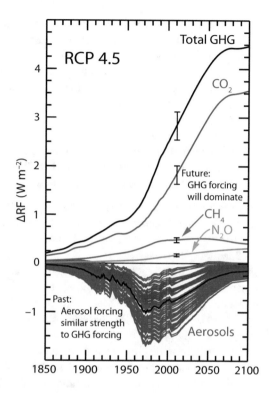

1.3 Methods

Most of the figures are composites formed by combining data from publicly available data archives. Here we provide details on webpage addresses of these archives, citations to the scientific papers that describe the measurements, as well as details regarding how the data has been processed. Electronic copies of the figures are available on-line at http://parisbeaconofhope.org.

Figure 1.1 shows estimates of the global mean surface temperature anomaly (ΔT) relative to the pre-industrial baseline and the mixing ratio of atmospheric CO_2, plotted using a logarithmic scale. The figure is broken up into six intervals, denoted using Era. Sources of ΔT and CO_2 for each Era are described below.

Era 1, ΔT is based on two data records:

(i) 1850 to present: the HadCRUT4.4.0.0 global, annual mean temperature record based on thermometer measurements, provided by the Climatic Research Unit (CRU) of the University of East Anglia, in conjunction with the Hadley Centre of the United Kingdom Met Office (Jones et al. 2012), archived at:
http://www.metoffice.gov.uk/hadobs/hadcrut4/data/4.4.0.0/time_series/
HadCRUT.4.4.0.0.annual_ns_avg.txt
Column 2 of this file tabulates ΔT relative to their 1961–1990 baseline. We have added 0.3134 °C to each data point, in order to place the measurements on the

1850–1900 baseline used throughout this book (i.e., the mean value of ΔT after this adjustment, averaged over years 1850–1900, is by definition zero).

(ii) 1000 ybp to 1849: a temperature reconstruction based on various proxies, such as tree rings, corals, etc. published by Jones and Mann (2004), archived by the National Centers for Environmental Information (NCEI) of the US National Oceanographic and Atmospheric Administration (NOAA) at:

ftp://ftp.ncdc.noaa.gov/pub/data/paleo/contributions_by_author/jones2004/jonesmannrogfig5.txt

Column 6 of this file tabulates global ΔT smoothed with a low pass filter, relative to their 1856–1980 baseline; 0.2657 °C has been added to place the measurements on our 1850–1900 baseline.

Eras 2 and 3, ΔT is based on the European Project for Ice Coring in Antarctica (EPICA) Dome C record (Jouzel et al. 2007) archived at:

http://www1.ncdc.noaa.gov/pub/data/paleo/icecore/antarctica/epica_domec/edc3deuttemp2007.txt

This record is based on analysis of the isotopic composition of the ice core, which is sensitive to temperature conditions at the time the ice formed. Two adjustments have been applied. First, we have subtracted 0.4250 °C from each data point to place the record on our 1850–1900 baseline. Second, since the ice core record represents temperature anomalies in Antarctica, which are larger than for other parts of the world, we have multiplied each data point by 0.463 to account for this difference. This multiplicative factor, based on analysis of the relation between Arctic and global warming over the modern time period (Chylek and Lohmann 2005), is in good agreement with the climate model simulations of the relation between warming in Antarctica and throughout the world (Masson-Delmotte et al. 2006).

Eras 4 and 5, ΔT is based on changes in Earth's surface temperature inferred from observations of isotopic composition of the shells preserved in deep seas cores (Hansen et al. 2013), archived at Columbia University:

http://www.columbia.edu/~mhs119/Sensitivity+SL+CO2/Table.txt

Data in column 6, labeled Ts, are used. The authors have related these deep sea core inferences to 14 °C, which is the globally averaged surface temperature from 1961 to 1990. We have subtracted 14 °C from each data point to turn the record into an anomaly relative the 1961–1990 baseline, then added 0.3134 °C to each data point to place this record on our 1850–1900 baseline.

Era6, ΔT is based on the isotopic composition of marine carbonates corrected for the influence of oceanic acidity and adjusted also for modeled variations of ancient, atmospheric CO_2 (Royer et al. 2004) archived at:

http://www.realclimate.org/docs/Temp-summary-from-Royer-et-al-2004.xls

Data in column D of this file were used. The authors have estimated changes in deep sea temperature relative to present. We have converted to surface temperature anomaly by multiplying their record by 2.5, the ratio of changes in global surface temperature to deep sea temperature according to equation 4.2 of Hansen et al. (2013). We have interpreted present to mean the 1961–1990 baseline, so we have also added 0.3134 °C to the data so that this record is also reflective of our 1850–1900 baseline.

Era 1, CO_2 is based on three data records:

(i) 1980 to present: global, annual average CO_2 provided by the NOAA Earth System Research Laboratory (ESRL) (Ballantyne et al. 2012) at:
     ftp://aftp.cmdl.noaa.gov/products/trends/co2/co2_annmean_gl.txt

(ii) 1765–1979: global, annual average CO_2 provided by the Potsdam Institute for Climate Research, developed as model inputs (Meinshausen et al. 2011) for climate model simulations used in the 2013 IPCC report, archived at:
     http://www.pik-potsdam.de/~mmalte/rcps/data/RCP45_MIDYEAR_CONCENTRATIONS.DAT
     This web address points to the RCP4.5 scenario, which is featured heavily in Chaps. 2 and 3 of this book. Since the record of CO_2 over this time period is constrained by observations, the numerical values of CO_2 for 1765–1979 are identical for all four RCP scenarios used in IPCC (2013).

(iii) 1000 ybp to 1764: The Law Dome Ice Core of record CO_2 (MacFarling Meure et al. 2006) archived by NOAA National Centers for Environmental Information (NECI) at:
     http://www1.ncdc.noaa.gov/pub/data/paleo/icecore/antarctica/law/law2006.txt
     This record is based on laboratory measurement of the CO_2 content of air pockets extracted from the upper part of the ice core, termed the firn layer.

Eras 2 and 3, CO_2 is based on a merged ice core data set that combines measurements from seven ice cores archived at:
   http://www1.ncdc.noaa.gov/pub/data/paleo/icecore/antarctica/antarctica2015co2.xls
This record is also based on laboratory measurement of CO_2 in air extracted from the ice (e.g., Petit et al. 1999). Column 2 of the CO2_Composite tab of the Excel file has been used; this composite is based on ten publications, all cited in the file.

Eras 4, 5, and 6, CO_2 is based on proxy estimates from five methods, originating from more than a hundred individual publications, summarized by Royer et al. (2012) and Peppe and Royer (2015). We have used a data file containing these observations sent to us by Dana Royer, senior author of these papers. Data for each proxy was first averaged, for all points falling within temporal bins of width 1 million years for Era 4, 5 million years for Era 5, and 50 million years for Era 5. Then, for each time bin, all available proxy means were averaged, resulting in the CO_2 time series connected by the blue lines. The error bars represent the minimum and maximum of the various proxy means available for specific time intervals. If CO_2 from a proxy was not available for a particular bin, a linear interpolation across adjacent time bins was applied, if possible. Otherwise, CO_2 from that missing proxy was treated as not available. The time ranges spanned by the five proxies are: paleosols (1–400 Mybp); alkenones (1–40 Mybp); stomata (1–400 Mypb); boron (1–15 and 35–55 Mybp); and liverworts (50–200 Mypb). Finally, the paleosol record as corrected by Breecker et al. (2009) was used.

Figure 1.2 shows values for the global mean surface temperature anomaly (ΔT) relative to the 1850–1900 baseline from two sources. For years prior to 1855, the proxy temperature time series of Jones and Mann (2004) was used. For 1855 to present, instrument data from HadCRUT.4.4.0.0 (Jones et al. 2012) was used. Both datasets were downloaded from the websites described for Fig. 1.1. A 21-year running mean was used to smooth HadCRUT.4.4.0.0 record up to 2008; data from 2009 to 2015 represent unsmoothed annual averages.

The GHG data in Fig. 1.2 is based on three data records. For 0 AD to 1764, observations of CO_2, CH_4, and N_2O are based on the Law Dome Ice Core (MacFarling Meure et al. 2006) archived by NOAA NECI at:

http://www1.ncdc.noaa.gov/pub/data/paleo/icecore/antarctica/law/law2006.txt
For 1765 to modern times (1979 for CO_2; 1983 for CH_4, 1977 for N_2O), GHG abundances are based on the RCP 4.5 archive at:

http://www.pik-potsdam.de/~mmalte/rcps/data/RCP45_MIDYEAR_CONCENTRATIONS.DAT
For years since 1980 for CO_2, 1984 for CH_4, and 1978 for N_2O, GHG abundances are based on observations provided by NOAA ESRL (Ballantyne et al. 2012; Dlugokencky et al. 2009; Montzka et al. 2011) at:

ftp://aftp.cmdl.noaa.gov/products/trends/co2/co2_annmean_gl.txt
ftp://aftp.cmdl.noaa.gov/products/trends/ch4/ch4_annmean_gl.txt
ftp://ftp.cmdl.noaa.gov/hats/n2o/combined/HATS_global_N2O.txt

Population data in Fig. 1.2, for years up to and including 1950, are from the History Database of the Global Environment (HYDE) of the Netherlands Environmental Assessment Agency (Klein Goldewijk et al. 2010). For 1951 to present, population data from the Department of Economic and Social Affairs of the United Nations (United Nations 2015) have been used. The population databases are maintained at:

http://themasites.pbl.nl/tridion/en/themasites/hyde/index.html
http://esa.un.org/unpd/wpp/Download/Standard/Population

Figure 1.3 shows values of RF forcing of climate, relative to year 1765, for the RCP 4.5 scenario (Meinshausen et al. 2011) from file:

http://www.pik-potsdam.de/~mmalte/rcps/data/RCP45_MIDYEAR_RADFORCING.DAT
maintained at the Potsdam Institute for Climate Research. The Anthropogenic Aerosols terms includes the direct radiative effect of aerosols, the perturbation to the reflectivity of clouds induced by aerosols, and the darkening of snow caused by the deposition of black carbon. The following types of aerosols were considered: sulfate, organic carbon and black carbon from both fossil fuel combustion and biomass burning, nitrate, and mineral dust. The total anthropogenic term combines the contributions to RF of climate from all GHGs released by human activity, plus RF of climate due to aerosols, depletion of stratospheric O_3, the increase of tropospheric O_3, and rising surface reflectivity due to land use change. Figure 1.3 also shows the global mean surface temperature anomaly (ΔT) relative to pre-industrial (1850–1900) baseline. Data sources for ΔT are the same as for Fig. 1.2.

Figure 1.4 shows the change in the radiative forcing (ΔRF) over the course of the Anthropocene (in this case, 1750–2011) from Chap. 8 of IPCC (2013). Numerical estimates for ΔRF are shown when available; otherwise, numerical estimates for the change in Effective Radiative Forcing (ΔERF) are used. Effective Radiative Forcing (ERF) is a new concept introduced in IPCC (2013), based on model simulations that

Table 1.3 ΔRF values used in Fig. 1.4

Term	ΔRF (W m⁻²)	Range of ΔRF (W m⁻²)	Origin within Chap. 8 of IPCC (2013)
CO_2	1.82	1.63–2.01	Table 8.2, RF
CH_4	0.48	0.43–0.53	Table 8.2, RF
N_2O	0.17	0.14–0.20	Table 8.2, RF
ODS[a]	0.33	0.297–0.363	Table 8.2, RF
Other F-Gases[b]	0.03	0.027–0.033	Table 8.2, RF
Tropospheric O_3	0.4	0.2–0.6	Table 8.6, RF
Stratospheric O_3	−0.05	−0.15 to 0.05	Table 8.6, RF
Stratospheric H_2O[c]	0.07	0.02–0.12	Table 8.6, RF
Contrails and Contrail-Induced Cirrus	0.05	0.02–0.15	Table 8.6, ERF
Surface Reflectivity: Land Use Change	−0.15	−0.25 to −0.05	Table 8.6, RF
Surface Refl.: Black Carbon on Snow	0.04	0.02–0.09	Table 8.6, RF
Aerosol Direct Effect	−0.45	−0.95 to 0.05	Table 8.6, ERF
Aerosol-Cloud Interaction	−0.45	−1.2 to 0.0	Table 8.6, ERF
Total Anthropogenic	2.3	1.1–3.3	Table 8.6, ERF
Solar Irradiance	0.05	0.0–0.10	Table 8.6, RF

[a]The definition of Ozone Depleting Substances used in Chap. 8 of IPCC (2013) combines the RF of climate due to CFC-11, CFC-12, CFC-13, CFC-113, CFC-114, CFC-115, HCFC-141b, HCFC-142b, CH_3CCl_3, CCl_4, Halon-1211, and Halon-1301. The IPCC (2013) definition appears to neglect Halon-1202, Halon-2402, CH_3Cl, and CH_3Br. The ΔRF of these four compounds is quite small, less than 0.002 W m⁻², so Fig. 1.4 would look identical had these four gases been considered
[b]This term considers the RF of climate due to HFCs, PFCs, SF_6, and a few other long-lived fluorinated species. The IPCC (2013) definition combines the RF of climate due to HFC-23, HFC-32, HFC-125, HFC-134a, HFC-143a, HFC-152a, CF_4, C_2F_6, SF_6, SO_2F_6, and NF_3
[c]This term represents the RF of climate due to the increase in stratospheric H_2O driven by rising levels of tropospheric CH_4. It does not include radiative effects of changes in stratospheric H_2O that occur in response to global warming (Solomon et al. 2010)

allow physical variables within the troposphere to respond to perturbations, except for those ocean and sea ice variables. For computations of RF, all surface and tropospheric conditions are kept fixed. Quoting Box 8.1 of IPCC (2013), "the calculation of ERF requires longer simulations with more complex models than the calculation of RF, but the inclusion of the additional rapid adjustments makes ERF a better indicator of the eventual global mean temperature response, especially for aerosols". We have used a mixture of ΔRF and ΔERF values for Fig. 1.4 because this is all that is available from Chap. 8 of IPCC (2013). Table 1.3 provides numerical estimates of the value, uncertainty, and origin of the data used in Fig. 1.4. All uncertainties represent 5–95 % confidence intervals and are given as a range, rather than a plus and minus value, since some are asymmetric about the mean.

Figure 1.5 shows a profile of the change in temperature over the time period 1959–2012, based on radiosonde observations collected in the latitude range 30°S to 30°N (Sherwood and Nishant 2015). Data reflect the Iterative Universal Kriging

(IUKv2) processing. Results are displayed as a function of altitude, rather than pressure, using a standard climatology for altitude versus pressure of the tropical atmosphere. The tropopause has been placed at the altitude corresponding to a pressure of 100 hPa. The patterns of tropospheric warming, stratospheric cooling, and drop in the tropospheric lapse rate (i.e., more warming aloft than at the surface) illustrated in Fig. 1.5 are seen throughout the global atmosphere, in addition to the tropics (Sherwood and Nishant 2015).

Figure 1.6 shows CO_2 from Mauna Loa Observatory (Keeling et al. 1976) and global annual average CO_2 (Ballantyne et al. 2012) provided by NOAA ESRL at:

ftp://ftp.cmdl.noaa.gov/products/trends/co2/co2_mm_mlo.txt

ftp://ftp.cmdl.noaa.gov/products/trends/co2/co2_annmean_gl.txt

The global CO_2 record given at the above URL starts in 1980. We have extended this record back to 1959 using annual, global average CO_2 growth rates given at:

http://www.esrl.noaa.gov/gmd/ccgg/trends/global.html#global_growth

The data used to construct the CO_2 emissions from the combustion of fossil fuel (Boden et al. 2013) plus land use change (Houghton et al. 2012) (green bars, Fig. 1.6a) originate from file:

http://cdiac.ornl.gov/ftp/Global_Carbon_Project/Global_Carbon_Budget_2015_v1.1.xlsx

hosted by the Carbon Dioxide Information Analysis Center (CDIAC) at the US Department of Energy's (DOE) Oak Ridge National Laboratory (ORNL). This same file is also provided by the Global Carbon Budget at:

http://www.globalcarbonproject.org/carbonbudget/15/files/Global_Carbon_Budget_2015v1.1.xlsx

Contents of this file, which contains much more information than used here, are described by Le Quéré et al. (2015). The blue bars are found by multiplying the difference in annual average CO_2 mixing ratio, units of ppm, by 7.768, to arrive at the mass of CO_2 in Gt (see Le Quéré et al. (2015)). Finally, the Tropical Pacific ENSO index represents the anomaly of sea surface temperature in the region bounded by 20°S to 20°N latitude and 160°E to 80°W longitude, relative to a long-term climatology. Monthly values of this index have been computed as described by Zhang et al. (1997), using HadSST3.1.1.0 sea surface temperature data (Kennedy et al. 2011a, b) provided by the Hadley Centre of the United Kingdom Met Office in file:

http://hadobs.metoffice.com/hadsst3/data/HadSST.3.1.1.0/netcdf/HadSST.3.1.1.0.median_netcdf.zip

Figure 1.7 shows CO_2 (Keeling et al. 1976), the O_2/N_2 ratio (Keeling et al. 1996), and $\delta^{13}C$ of CO_2 (Keeling et al. 2005) from Mauna Loa Observatory (MLO) as well as CO_2 from the South Pole (SPO) (Tans et al. 1990). For CO_2, the solid black line shows monthly mean data from NOAA ESRL, based on the same file given in Methods for Fig. 1.6. Daily measurements of CO_2 at MLO (dots) are based on data provided by the Scripps Institution of Oceanography (SIO) of the University of California, San Diego at:

http://scrippsco2.ucsd.edu/assets/data/atmospheric/stations/flask_co2/daily/daily_flask_co2_mlo.csv

Monthly mean CO_2 at SPO (red line) is based on data provided by NOAA ESRL at:

http://scrippsco2.ucsd.edu/assets/data/atmospheric/stations/flask_co2/monthly/monthly_flask_co2_spo.csv

Monthly (line) and daily (dots) observations of the O_2/N_2 ratio are based on data archived by SIO at:

http://scrippso2.ucsd.edu/sites/default/files/data/o2_data/o2_monthly/mloo.txt

http://scrippso2.ucsd.edu/sites/default/files/data/o2_data/o2_daily/mlooav.csv

Finally, monthly (line) and daily (dots) $^{13}\delta CO_2$ is also based on data from SIO, at:

http://scrippsco2.ucsd.edu/assets/data/atmospheric/stations/flask_isotopic/monthly/monthly_flask_c13_mlo.csv

http://scrippsco2.ucsd.edu/assets/data/atmospheric/stations/flask_isotopic/daily/daily_flask_c13_mlo.csv

Figure 1.8 shows the difference between annual average CO_2 at MLO and CO_2 at SPO versus total anthropogenic emissions of CO_2. Here, we formed annual average CO_2 for each station from monthly mean values, based on in situ and flask sampling archived by SIO at:

http://scrippsco2.ucsd.edu/assets/data/atmospheric/stations/in_situ_co2/monthly/monthly_in_situ_co2_mlo.csv

http://scrippsco2.ucsd.edu/assets/data/atmospheric/stations/flask_co2/monthly/monthly_flask_co2_spo.csv

The total anthropogenic emissions of CO_2 are annual tabulations, reflecting the sum of combustion of fossil fuel (Boden et al. 2013) plus land use change (Houghton et al. 2012), and originate from the Global Carbon Project archive described in Methods for Fig. 1.6.

Figure 1.9 shows an estimate for the sources and sinks of atmospheric CH_4. Numerical values of the CH_4 source for total humans (335 Tg year^{-1}), total natural (218 Tg year^{-1}), natural wetlands (175 Tg year^{-1}), other natural (43 Tg year^{-1}) are from the top-down estimates for 2000–2009 given in Table 1 of Kirschke et al. (2013). The apportionment of the human sources is based on Fig. 1 of Conrad (2009). Numerical values of the total sink (550 Tg year^{-1}) and the sink due to soils (32 Tg year^{-1}) are also based on the top-down estimates from table 1 of Kirschke et al. (2013), whereas the sinks due to chemical loss via reactions with tropospheric OH (452.8 Tg year^{-1}), reactions with tropospheric Cl (21.4 Tg year^{-1}), and stratospheric chemistry (43.7 Tg year^{-1}) are based by scaling the bottom-up estimates of these quantities given in Table 1 of Kirschke et al. (2013) for the decade 2000–2009 by the ratio $518/604 = 0.8576$, where 518 Tg year^{-1} is the total chemical sink found using the trop-down approach and 604 Tg year^{-1} is the total chemical sink found using the bottom-up approach. The numbers have been combined in this manner to provide a self-consistent estimate of the CH_4 source and sink terms by combining best available information from several studies and approaches.

Figure 1.10 shows ΔRF due to CO_2, CH_4, N_2O, and all anthropogenic GHGs from the RCP 4.5 scenario (Meinshausen et al. 2011). The numerical values have been obtained from the same file used to find ΔRF for Fig. 1.3. The figure also shows 71 plausible scenarios for total ΔRF due to anthropogenic aerosols from Smith and Bond (2014). A data file containing the 71 time series for ΔRF of climate due to aerosols that appeared as Fig. 1.4 of this paper was sent to us by Steven J. Smith, corresponding author of this paper.

References

Anderson DC, Nicely JM, Salawitch RJ, Canty TP, Dickerson RR, Hanisco TF, Wolfe GM, Apel EC, Atlas E, Bannan T, Bauguitte S, Blake NJ, Bresch JF, Campos TL, Carpenter LJ, Cohen MD, Evans M, Fernandez RP, Kahn BH, Kinnison DE, Hall SR, Harris NRP, Hornbrook RS, Lamarque J-F, Le Breton M, Lee JD, Percival C, Pfister L, Pierce RB, Riemer DD, Saiz-Lopez A, Stunder BJB, Thompson AM, Ullmann K, Vaughan A, Weinheimer AJ (2016) A pervasive role for biomass burning in tropical high ozone/low water structures. Nat Commun 7, 10267. doi:10.1038/ncomms10267

Arnold T, Mühle J, Salameh PK, Harth CM, Ivy DJ, Weiss RF (2012) Automated measurement of nitrogen trifluoride in ambient air. Anal Chem 84(11):4798–4804. doi:10.1021/ac300373e

Ballantyne AP, Alden CB, Miller JB, Tans PP, White JWC (2012) Increase in observed net carbon dioxide uptake by land and oceans during the past 50 years. Nature 488(7409):70–72

Bard E, Raisbeck G, Yiou F, Jouzel J (2000) Solar irradiance during the last 1200 years based on cosmogenic nuclides. Tellus B 52(3):985–992. doi:10.1034/j.1600-0889.2000.d01-7.x

Barnola JM, Raynaud D, Korotkevich YS, Lorius C (1987) Vostok ice core provides 160,000-year record of atmospheric CO_2. Nature 329(6138):408–414

Bell ML, Dominici F, Keita E, Zeger SL, Samet JM (2007) Spatial and temporal variation in $PM_{2.5}$ chemical composition in the United States for health effects studies. Environ Health Perspect 115(7):989–995

Bera PP, Francisco JS, Lee TJ (2009) Identifying the molecular origin of global warming. J Phys Chem A 113(45):12694–12699. doi:10.1021/jp905097g

Berner RA (1997) The rise of plants and their effect on weathering and atmospheric CO_2. Science 276(5312):544–546. doi:10.1126/science.276.5312.544

Berner RA, Lasaga AC, Garrels RM (1983) The carbonate-silicate geochemical cycle and its effect on atmospheric carbon dioxide over the past 100 million years. Am J Sci 283(7):641–683. doi:10.2475/ajs.283.7.641

Boden TA, Marland G, Andres RJ (2013) Global, regional, and national fossil-fuel CO_2 emissions. doi:10.3334/CDIAC/00001_V2013

Bond TC, Doherty SJ, Fahey DW, Forster PM, Berntsen T, DeAngelo BJ, Flanner MG, Ghan S, Kärcher B, Koch D, Kinne S, Kondo Y, Quinn PK, Sarofim MC, Schultz MG, Schulz M, Venkataraman C, Zhang H, Zhang S, Bellouin N, Guttikunda SK, Hopke PK, Jacobson MZ, Kaiser JW, Klimont Z, Lohmann U, Schwarz JP, Shindell D, Storelvmo T, Warren SG, Zender CS (2013) Bounding the role of black carbon in the climate system: a scientific assessment. J Geophys Res Atmos 118(11):5380–5552. doi:10.1002/jgrd.50171

Bony S, Colman R, Kattsov VM, Allan RP, Bretherton CS, Dufresne J-L, Hall A, Hallegatte S, Holland MM, Ingram W, Randall DA, Doden BJ, Tselioudis G, Webb MJ (2006) How well do we understand and evaluate climate change feedback processes? J Clim 19:3445–3482. doi:10.1029/2005GL023851

Bousquet P, Peylin P, Ciais P, Le Quéré C, Friedlingstein P, Tans PP (2000) Regional changes in carbon dioxide fluxes of land and oceans since 1980. Science 290(5495):1342–1346. doi:10.1126/science.290.5495.1342

Brasier MD, Green OR, Jephcoat AP, Kleppe AK, Van Kranendonk MJ, Lindsay JF, Steele A, Grassineau NV (2002) Questioning the evidence for Earth's oldest fossils. Nature 416(6876):76–81. doi:10.1038/416076a

Breecker DO, Sharp ZD, McFadden LD (2009) Seasonal bias in the formation and stable isotopic composition of pedogenic carbonate in modern soils from central New Mexico, USA. GSA Bull 121(3–4):630–640. doi:10.1130/b26413.1

Brook EJ, Harder S, Severinghaus J, Steig EJ, Sucher CM (2000) On the origin and timing of rapid changes in atmospheric methane during the Last Glacial Period. Glob Biogeochem Cycles 14(2):559–572. doi:10.1029/1999GB001182

Caillon N, Severinghaus JP, Jouzel J, Barnola J-M, Kang J, Lipenkov VY (2003) Timing of atmospheric CO_2 and Antarctic temperature changes across termination III. Science 299(5613):1728–1731. doi:10.1126/science.1078758

Canfield DE, Raiswell R (1999) The evolution of the sulfur cycle. Am J Sci 299(7–9):697–723. doi:10.2475/ajs.299.7-9.697

Carslaw KS, Lee LA, Reddington CL, Pringle KJ, Rap A, Forster PM, Mann GW, Spracklen DV, Woodhouse MT, Regayre LA, Pierce JR (2013) Large contribution of natural aerosols to uncertainty in indirect forcing. Nature 503(7474):67–71. doi:10.1038/nature12674

Carto SL, Weaver AJ, Hetherington R, Lam Y, Wiebe EC (2009) Out of Africa and into an ice age: on the role of global climate change in the late Pleistocene migration of early modern humans out of Africa. J Hum Evol 56(2):139–151. doi:10.1016/j.jhevol.2008.09.004

Chylek P, Lohmann U (2005) Ratio of the Greenland to global temperature change: comparison of observations and climate modeling results. Geophys Res Lett 32(14), L14705. doi:10.1029/2005GL023552

Chylek P, Lohmann U (2008) Aerosol radiative forcing and climate sensitivity deduced from the Last Glacial Maximum to Holocene transition. Geophys Res Lett 35(4), L23703. doi:10.1029/2007GL032759

Conrad R (2009) The global methane cycle: recent advances in understanding the microbial processes involved. Environ Microbiol Rep 1(5):285–292. doi:10.1111/j.1758-2229.2009.00038.x

Crutzen PJ, Stoermer EF (2000) The Anthropocene. IGBP Newsletter, Royal Swedish Academy of Sciences, vol 41

Crutzen PJ, Mosier AR, Smith KA, Winiwarter W (2016) N_2O release from agro-biofuel production negates global warming reduction by replacing fossil fuels. In: Crutzen JP, Brauch GH (eds) Paul J. Crutzen: a pioneer on atmospheric chemistry and climate change in the anthropocene. Springer International Publishing, Cham, pp 227–238

Dessert C, Dupré B, François LM, Schott J, Gaillardet J, Chakrapani G, Bajpai S (2001) Erosion of Deccan Traps determined by river geochemistry: impact on the global climate and the 87Sr/86Sr ratio of seawater. Earth Planet Sci Lett 188(3–4):459–474, http://dx.doi.org/10.1016/S0012-821X(01)00317-X

Dlugokencky EJ, Bruhwiler L, White JWC, Emmons LK, Novelli PC, Montzka SA, Masarie KA, Lang PM, Crotwell AM, Miller JB, Gatti LV (2009) Observational constraints on recent increases in the atmospheric CH_4 burden. Geophys Res Lett 36(18), L18803. doi:10.1029/2009GL039780

Dominici F, Peng RD, Bell ML, Pham L, McDermott A, Zeger SL, Samet JM (2006) Fine particulate air pollution and hospital admission for cardiovascular and respiratory diseases. JAMA 295(10):1127–1134. doi:10.1001/jama.295.10.1127

Erb MP, Broccoli AJ, Clement AC (2013) The contribution of radiative feedbacks to orbitally driven climate change. J Clim 26(16):5897–5914. doi:10.1175/JCLI-D-12-00419.1

Fan S-M, Blaine TL, Sarmiento JL (1999) Terrestrial carbon sink in the Northern Hemisphere estimated from the atmospheric CO_2 difference between Mauna Loa and the South Pole since 1959. Tellus B 51(5):863–870. doi:10.1034/j.1600-0889.1999.t01-4-00001.x

Frei R, Gaucher C, Poulton SW, Canfield DE (2009) Fluctuations in Precambrian atmospheric oxygenation recorded by chromium isotopes. Nature 461(7261):250–253. doi:10.1038/nature08266

Galloway JN, Winiwarter W, Leip A, Leach AM, Bleeker A, Erisman JW (2014) Nitrogen footprints: past, present and future. Environ Res Lett 9(11):115003

Gepts P, Papa R (2001) Evolution during domestication. In: Encyclopedia of life sciences. Wiley, New York. doi:10.1038/npg.els.0003071

Gerlach T (2011) Volcanic versus anthropogenic carbon dioxide. Eos 92(24):201–208

Hansen J, Sato M, Russell G, Kharecha P (2013) Climate sensitivity, sea level and atmospheric carbon dioxide. Philos Trans R Soc Lond A Math Phys Eng Sci 371(2001). doi:10.1098/rsta.2012.0294

Hay WW, Barron EJ, Thompson SL (1990) Global atmospheric circulation experiments on an Earth with polar and tropical continents. J Geol Soc 147(5):749–757. doi:10.1144/gsjgs.147.5.0749

He H, Vinnikov KY, Li C, Krotkov NA, Jongeward AR, Li Z, Stehr JW, Hains JC, Dickerson RR (2016) Response of SO_2 and particulate air pollution to local and regional emission controls: a case study in Maryland. Earth's Future 4(4):94–109. doi:10.1002/2015EF000330

Houghton RA, House JI, Pongratz J, van der Werf GR, DeFries RS, Hansen MC, Le Quéré C, Ramankutty N (2012) Carbon emissions from land use and land-cover change. Biogeosciences 9(12):5125–5142. doi:10.5194/bg-9-5125-2012

Imbrie J, Imbrie KP (1979) Ice ages: solving the mystery. Harvard University Press, Cambridge, MA

IPCC (1995) Climate change 1995: the science of climate change. Contribution of working group I to the second assessment report of the intergovernmental panel on climate change. Cambridge, UK and New York, NY, USA

IPCC (2001) Climate change 2001: the scientific basis. contribution of working group I to the third assessment report of the intergovernmental panel on climate change. Cambridge, UK and New York, NY, USA

IPCC (2007) Climate change 2007: the physical science basis. Contribution of working group I to the fourth assessment report of the intergovernmental panel on climate change. Cambridge, UK and New York, NY, USA

IPCC (2013) Climate change 2013: the physical science basis. Contribution of working group I to the fifth assessment report of the intergovernmental panel on climate change. Cambridge, UK and New York, NY, USA

IPCC/TEAP (2005) Safeguarding the ozone layer and the global climate systems: issues related to hydrofluorocarbons and perfluorocarbons. Prepared by Working Group I and III of the Intergovernmental Panel on Climate Change, and the Technology and Economic Assessment Panel. Cambridge, UK and New York, NY, USA

Jimenez JL, Canagaratna MR, Donahue NM, Prevot AS, Zhang Q, Kroll JH, DeCarlo PF, Allan JD, Coe H, Ng NL, Aiken AC, Docherty KS, Ulbrich IM, Grieshop AP, Robinson AL, Duplissy J, Smith JD, Wilson KR, Lanz VA, Hueglin C, Sun YL, Tian J, Laaksonen A, Raatikainen T, Rautiainen J, Vaattovaara P, Ehn M, Kulmala M, Tomlinson JM, Collins DR, Cubison MJ, Dunlea EJ, Huffman JA, Onasch TB, Alfarra MR, Williams PI, Bower K, Kondo Y, Schneider J, Drewnick F, Borrmann S, Weimer S, Demerjian K, Salcedo D, Cottrell L, Griffin R, Takami A, Miyoshi T, Hatakeyama S, Shimono A, Sun JY, Zhang YM, Dzepina K, Kimmel JR, Sueper D, Jayne JT, Herndon SC, Trimborn AM, Williams LR, Wood EC, Middlebrook AM, Kolb CE, Baltensperger U, Worsnop DR (2009) Evolution of organic aerosols in the atmosphere. Science 326(5959):1525–1529. doi:10.1126/science.1180353

Johanson D, White T (1979) A systematic assessment of early African hominids. Science 203(4378):321–330. doi:10.1126/science.104384

Jones PD, Mann ME (2004) Climate over past millennia. Rev Geophys 42(2):RG2002. doi:10.1029/2003RG000143

Jones PD, Lister DH, Osborn TJ, Harpham C, Salmon M, Morice CP (2012) Hemispheric and large-scale land-surface air temperature variations: an extensive revision and an update to 2010. J Geophys Res 117(D5):D05127. doi:10.1029/2011jd017139

Jouzel J, Lorius C, Petit JR, Genthon C, Barkov NI, Kotlyakov VM, Petrov VM (1987) Vostok ice core: a continuous isotope temperature record over the last climatic cycle (160,000 years). Nature 329(6138):403–408

Jouzel J, Masson-Delmotte V, Cattani O, Dreyfus G, Falourd S, Hoffmann G, Minster B, Nouet J, Barnola JM, Chappellaz J, Fischer H, Gallet JC, Johnsen S, Leuenberger M, Loulergue L, Luethi D, Oerter H, Parrenin F, Raisbeck G, Raynaud D, Schilt A, Schwander J, Selmo E, Souchez R, Spahni R, Stauffer B, Steffensen JP, Stenni B, Stocker TF, Tison JL, Werner M,

Wolff EW (2007) Orbital and millennial Antarctic climate variability over the past 800,000 years. Science 317(5839):793–796. doi:10.1126/science.1141038

Kahn RA (2012) Reducing the uncertainties in direct aerosol radiative forcing. Surv Geophys 33(3):701–721. doi:10.1007/s10712-011-9153-z

Kasting JF, Siefert JL (2002) Life and the evolution of Earth's atmosphere. Science 296(5570):1066–1068. doi:10.1126/science.1071184

Keeling CD, Bacastow RB, Bainbridge AE, Ekdahl CA, Guenther PR, Waterman LS, Chin JFS (1976) Atmospheric carbon dioxide variations at Mauna Loa Observatory, Hawaii. Tellus 28(6):538–551. doi:10.1111/j.2153-3490.1976.tb00701.x

Keeling RF, Piper SC, Heimann M (1996) Global and hemispheric CO_2 sinks deduced from changes in atmospheric O_2 concentration. Nature 381(6579):218–221

Keeling CD, Piper SC, Bacastow RB, Wahlen M, Whorf TP, Heimann M, Meijer HA (2005) Atmospheric CO_2 and $^{13}CO_2$ exchange with the terrestrial biosphere and oceans from 1978 to 2000: observations and carbon cycle implications. In: Baldwin IT, Caldwell MM, Heldmaier G et al (eds) A history of atmospheric CO_2 and its effects on plants, animals, and ecosystems. Springer, New York, pp 83–113. doi:10.1007/0-387-27048-5_5

Kennedy JJ, Rayner NA, Smith RO, Parker DE, Saunby M (2011a) Reassessing biases and other uncertainties in sea-surface temperature observations measured in situ since 1850, Part 1: Measurement and sampling uncertainties. J Geophys Res 116:D14103. doi:10.1029/2010JD015218

Kennedy JJ, Rayner NA, Smith RO, Parker DE, Saunby M (2011b) Reassessing biases and other uncertainties in sea-surface temperature observations measured in situ since 1850, Part 2: Biases and homogenisation. J Geophys Res 116, D14104. doi:10.1029/2010JD01522

Kenrick P, Crane PR (1997) The origin and early evolution of plants on land. Nature 389(6646):33–39

Keywood M, Kanakidou M, Stohl A, Dentener F, Grassi G, Meyer CP, Torseth K, Edwards D, Thompson AM, Lohmann U, Burrows J (2013) Fire in the air: biomass burning impacts in a changing climate. Crit Rev Environ Sci Technol 43(1):40–83. doi:10.1080/10643389.2011.604248

Kirschke S, Bousquet P, Ciais P, Saunois M, Canadell JG, Dlugokencky EJ, Bergamaschi P, Bergmann D, Blake DR, Bruhwiler L, Cameron-Smith P, Castaldi S, Chevallier F, Feng L, Fraser A, Heimann M, Hodson EL, Houweling S, Josse B, Fraser PJ, Krummel PB, Lamarque J-F, Langenfelds RL, Le Quere C, Naik V, O'Doherty S, Palmer PI, Pison I, Plummer D, Poulter B, Prinn RG, Rigby M, Ringeval B, Santini M, Schmidt M, Shindell DT, Simpson IJ, Spahni R, Steele LP, Strode SA, Sudo K, Szopa S, van der Werf GR, Voulgarakis A, van Weele M, Weiss RF, Williams JE, Zeng G (2013) Three decades of global methane sources and sinks. Nat Geosci 6(10):813–823. doi:10.1038/ngeo1955, http://www.nature.com/ngeo/journal/v6/n10/abs/ngeo1955.html#supplementary-information

Klein Goldewijk K, Beusen A, Janssen P (2010) Long-term dynamic modeling of global population and built-up area in a spatially explicit way: HYDE 3.1. The Holocene 20(4):565–573. doi:10.1177/0959683609356587

Koven CD, Ringeval B, Friedlingstein P, Ciais P, Cadule P, Khvorostyanov D, Krinner G, Tarnocai C (2011) Permafrost carbon-climate feedbacks accelerate global warming. Proc Natl Acad Sci 108(36):14769–14774. doi:10.1073/pnas.1103910108

Krotkov NA, McLinden CA, Li C, Lamsal LN, Celarier EA, Marchenko SV, Swartz WH, Bucsela EJ, Joiner J, Duncan BN, Boersma KF, Veefkind JP, Levelt PF, Fioletov VE, Dickerson RR, He H, Lu Z, Streets DG (2016) Aura OMI observations of regional SO_2 and NO_2 pollution changes from 2005 to 2015. Atmos Chem Phys 16(7):4605–4629. doi:10.5194/acp-16-4605-2016

Kvenvolden KA (1993) Gas hydrates—geological perspective and global change. Rev Geophys 31(2):173–187. doi:10.1029/93RG00268

Lacis AA, Mischenko MI (1995) Climate forcing, climate sensitivity, and climate response: a radiative modeling perspective on atmospheric aerosols. Aerosol forcing of climate: report of the Dahlem workshop on aerosol forcing of climate. Chichester, England/New York

Le Quéré C (2010) Trends in the land and ocean carbon uptake. Curr Opin Environ Sustain 2(4):219–224, http://dx.doi.org/10.1016/j.cosust.2010.06.003

Le Quéré C, Aumont O, Bopp L, Bousquet P, Ciais P, Francey R, Heimann M, Keeling CD, Keeling RF, Kheshgi H, Peylin P, Piper SC, Prentice IC, Rayner PJ (2003) Two decades of ocean CO_2 sink and variability. Tellus B 55(2):649–656. doi:10.1034/j.1600-0889.2003.00043.x

Le Quéré C, Moriarty R, Andrew RM, Canadell JG, Sitch S, Korsbakken JI, Friedlingstein P, Peters GP, Andres RJ, Boden TA, Houghton RA, House JI, Keeling RF, Tans P, Arneth A, Bakker DCE, Barbero L, Bopp L, Chang J, Chevallier F, Chini LP, Ciais P, Fader M, Feely RA, Gkritzalis T, Harris I, Hauck J, Ilyina T, Jain AK, Kato E, Kitidis V, Klein Goldewijk K, Koven C, Landschützer P, Lauvset SK, Lefèvre N, Lenton A, Lima ID, Metzl N, Millero F, Munro DR, Murata A, Nabel JEMS, Nakaoka S, Nojiri Y, O'Brien K, Olsen A, Ono T, Pérez FF, Pfeil B, Pierrot D, Poulter B, Rehder G, Rödenbeck C, Saito S, Schuster U, Schwinger J, Séférian R, Steinhoff T, Stocker BD, Sutton AJ, Takahashi T, Tilbrook B, van der Laan-Luijkx IT, van der Werf GR, van Heuven S, Vandemark D, Viovy N, Wiltshire A, Zaehle S, Zeng N (2015) Global carbon budget 2015. Earth Syst Sci Data 7(2):349–396. doi:10.5194/essd-7-349-2015

Loulergue L, Schilt A, Spahni R, Masson-Delmotte V, Blunier T, Lemieux B, Barnola J-M, Raynaud D, Stocker TF, Chappellaz J (2008) Orbital and millennial-scale features of atmospheric CH_4 over the past 800,000 years. Nature 453(7193):383–386, http://www.nature.com/nature/journal/v453/n7193/suppinfo/nature06950_S1.html

Lunt DJ, Foster GL, Haywood AM, Stone EJ (2008) Late Pliocene Greenland glaciation controlled by a decline in atmospheric CO_2 levels. Nature 454(7208):1102–1105. doi:10.1038/nature07223

MacFarling Meure C, Etheridge D, Trudinger C, Steele P, Langenfelds R, van Ommen T, Smith A, Elkins J (2006) Law Dome CO_2, CH_4 and N_2O ice core records extended to 2000 years BP. Geophys Res Lett 33(14). doi:10.1029/2006GL026152

Mann ME (2012) The hockey stick and the climate wars: dispatches from the front lines. Columbia University Press, Columbia

Mann ME, Bradley RS, Hughes MK (1999) Northern hemisphere temperatures during the past millennium: inferences, uncertainties, and limitations. Geophys Res Lett 26(6):759–762. doi:10.1029/1999GL900070

Margulis L, Sagan D (1986) Microcosmos: four billion years of evolution from our microbial ancestors. University of California Press, Berkeley, CA

Marino BD, McElroy MB, Salawitch RJ, Spaulding WG (1992) Glacial to interglacial variations in the carbon isotopic composition of atmospheric CO_2. Nature 357:461–466

Martin JH (1990) Glacial-interglacial CO_2 change: the iron hypothesis. Paleoceanography 5(1):1–13. doi:10.1029/PA005i001p00001

Marty B, Tolstikhin IN (1998) CO_2 fluxes from mid-ocean ridges, arcs and plumes. Chem Geol 145(3–4):233–248. http://dx.doi.org/10.1016/S0009-2541(97)00145-9

Mason BG, Pyle DM, Oppenheimer C (2004) The size and frequency of the largest explosive eruptions on Earth. Bull Volcanol 66(8):735–748. doi:10.1007/s00445-004-0355-9

Masson-Delmotte V, Kageyama M, Braconnot P, Charbit S, Krinner G, Ritz C, Guilyardi E, Jouzel J, Abe-Ouchi A, Crucifix M, Gladstone RM, Hewitt CD, Kitoh A, LeGrande AN, Marti O, Merkel U, Motoi T, Ohgaito R, Otto-Bliesner B, Peltier WR, Ross I, Valdes PJ, Vettoretti G, Weber SL, Wolk F, Yu Y (2006) Past and future polar amplification of climate change: climate model intercomparisons and ice-core constraints. Clim Dyn 26(5):513–529. doi:10.1007/s00382-005-0081-9

McLinden CA, Fioletov V, Shephard MW, Krotkov N, Li C, Martin RV, Moran MD, Joiner J (2016) Space-based detection of missing sulfur dioxide sources of global air pollution. Nat Geosci 9(7):496–500. doi:10.1038/ngeo2724, http://www.nature.com/ngeo/journal/vaop/ncurrent/abs/ngeo2724.html#supplementary-information

Meinshausen M, Smith SJ, Calvin K, Daniel JS, Kainuma MLT, Lamarque JF, Matsumoto K, Montzka SA, Raper SCB, Riahi K, Thomson A, Velders GJM, Vuuren DPP (2011) The RCP greenhouse gas concentrations and their extensions from 1765 to 2300. Clim Chang 109(1–2):213–241. doi:10.1007/s10584-011-0156-z

Minschwaner K, Salawitch R, McElroy M (1993) Absorption of solar radiation by O_2: implications for O_3 and lifetimes of N_2O, $CFCl_3$, and CF_2Cl_2. J Geophys Res 98(D6):10543–10561

Mitchell DM (2016) Attributing the forced components of observed stratospheric temperature variability to external drivers. Q J R Meteorol Soc 142(695):1041–1047. doi:10.1002/qj.2707

Moberg A, Sonechkin DM, Holmgren K, Datsenko NM, Karlen W (2005) Highly variable Northern Hemisphere temperatures reconstructed from low- and high-resolution proxy data. Nature 433(7026):613–617. doi:10.1038/nature03265

Montzka SA, Dlugokencky EJ, Butler JH (2011) Non-CO_2 greenhouse gases and climate change. Nature 476(7358):43–50

Morgan MG, Adams PJ, Keith DW (2006) Elicitation of expert judgments of aerosol forcing. Clim Chang 75(1–2):195–214. doi:10.1007/s10584-005-9025-y

Mühle J, Ganesan AL, Miller BR, Salameh PK, Harth CM, Greally BR, Rigby M, Porter LW, Steele LP, Trudinger CM, Krummel PB, O'Doherty S, Fraser PJ, Simmonds PG, Prinn RG, Weiss RF (2010) Perfluorocarbons in the global atmosphere: tetrafluoromethane, hexafluoroethane, and octafluoropropane. Atmos Chem Phys 10(11):5145–5164. doi:10.5194/acp-10-5145-2010

Myhre G (2009) Consistency between satellite-derived and modeled estimates of the direct aerosol effect. Science 325(5937):187–190. doi:10.1126/science.1174461

NAS (2006) Surface temperature reconstructions for the last 2,000 years. The National Academies Press, National Academy of Sciences

Newhall CG, Self S (1982) The volcanic explosivity index (VEI) an estimate of explosive magnitude for historical volcanism. J Geophys Res Oceans 87(C2):1231–1238. doi:10.1029/JC087iC02p01231

Parrenin F, Masson-Delmotte V, Köhler P, Raynaud D, Paillard D, Schwander J, Barbante C, Landais A, Wegner A, Jouzel J (2013) Synchronous change of atmospheric CO_2 and Antarctic temperature during the last deglacial warming. Science 339(6123):1060–1063. doi:10.1126/science.1226368

Peppe DJ, Royer DL (2015) Can climate feel the pressure? Science 348(6240):1210–1211. doi:10.1126/science.aac5264

Petit JR, Jouzel J, Raynaud D, Barkov NI, Barnola JM, Basile I, Bender M, Chappellaz J, Davis M, Delaygue G, Delmotte M, Kotlyakov VM, Legrand M, Lipenkov VY, Lorius C, Pepin L, Ritz C, Saltzman E, Stievenard M (1999) Climate and atmospheric history of the past 420,000 years from the Vostok ice core, Antarctica. Nature 399(6735):429–436. http://www.nature.com/nature/journal/v399/n6735/suppinfo/399429a0_S1.html

Pierrehumbert RT (2014) Short-lived climate pollution. Annu Rev Earth Planet Sci 42(1):341–379. doi:10.1146/annurev-earth-060313-054843

Prather MJ, Hsu J (2008) NF_3, the greenhouse gas missing from Kyoto. Geophys Res Lett 35(12), L12810. doi:10.1029/2008GL034542

Prather MJ, Holmes CD, Hsu J (2012) Reactive greenhouse gas scenarios: systematic exploration of uncertainties and the role of atmospheric chemistry. Geophys Res Lett 39(9). doi:10.1029/2012GL051440

Randerson JT, van der Werf GR, Collatz GJ, Giglio L, Still CJ, Kasibhatla P, Miller JB, White JWC, DeFries RS, Kasischke ES (2005) Fire emissions from C3 and C4 vegetation and their influence on interannual variability of atmospheric CO2 and δ13CO2. Global Biogeochem Cycles 19(2). doi:10.1029/2004GB002366

Ravishankara AR, Daniel JS, Portmann RW (2009) Nitrous oxide (N_2O): the dominant ozone-depleting substance emitted in the 21st century. Science 326(5949):123–125. doi:10.1126/science.1176985

Raymo ME, Ruddiman WF (1992) Tectonic forcing of late Cenozoic climate. Nature 359(6391):117–122

Revell LE, Tummon F, Salawitch RJ, Stenke A, Peter T (2015) The changing ozone depletion potential of N_2O in a future climate. Geophys Res Lett 42(22):10,047–10,055. doi:10.1002/2015GL065702

Revelle R, Suess HE (1957) Carbon dioxide exchange between atmosphere and ocean and the question of an increase of atmospheric CO_2 during the past decades. Tellus 9(1):18–27. doi:10.1111/j.2153-3490.1957.tb01849.x

Rigby M, Mühle J, Miller BR, Prinn RG, Krummel PB, Steele LP, Fraser PJ, Salameh PK, Harth CM, Weiss RF, Greally BR, O'Doherty S, Simmonds PG, Vollmer MK, Reimann S, Kim J, Kim KR, Wang HJ, Olivier JGJ, Dlugokencky EJ, Dutton GS, Hall BD, Elkins JW (2010) History of atmospheric SF_6 from 1973 to 2008. Atmos Chem Phys 10(21):10305–10320. doi:10.5194/acp-10-10305-2010

Rizzo AL, Jost H-J, Caracausi A, Paonita A, Liotta M, Martelli M (2014) Real-time measurements of the concentration and isotope composition of atmospheric and volcanic CO_2 at Mount Etna (Italy). Geophys Res Lett 41(7):2382–2389. doi:10.1002/2014GL059722

Rosenthal Y, Linsley BK, Oppo DW (2013) Pacific ocean heat content during the past 10,000 years. Science 342(6158):617–621. doi:10.1126/science.1240837

Royer DL, Berner RA, Montanez IP, Tabor NJ, Beerling DJ (2004) CO_2 as a primary driver of Phanerozoic climate. Geol Soc Am Today 14:4–10. doi:10.1130/1052-5173(2004)014<4:CAA PDO>2.0.CO;2

Royer DL, Pagani M, Beerling DJ (2012) Geobiological constraints on Earth system sensitivity to CO_2 during the Cretaceous and Cenozoic. Geobiology 10(4):298–310. doi:10.1111/j.1472-4669.2012.00320.x

Ruddiman WF (2003) The anthropogenic greenhouse era began thousands of years ago. Clim Chang 61(3):261–293. doi:10.1023/B:CLIM.0000004577.17928.fa

Sagan C, Mullen G (1972) Earth and mars: evolution of atmospheres and surface temperatures. Science 177(4043):52–56. doi:10.1126/science.177.4043.52

Santer BD, Painter JF, Bonfils C, Mears CA, Solomon S, Wigley TM, Gleckler PJ, Schmidt GA, Doutriaux C, Gillett NP, Taylor KE, Thorne PW, Wentz FJ (2013a) Human and natural influences on the changing thermal structure of the atmosphere. Proc Natl Acad Sci U S A 110(43):17235–17240. doi:10.1073/pnas.1305332110

Santer BD, Painter JF, Mears CA, Doutriaux C, Caldwell P, Arblaster JM, Cameron-Smith PJ, Gillett NP, Gleckler PJ, Lanzante J, Perlwitz J, Solomon S, Stott PA, Taylor KE, Terray L, Thorne PW, Wehner MF, Wentz FJ, Wigley TM, Wilcox LJ, Zou CZ (2013b) Identifying human influences on atmospheric temperature. Proc Natl Acad Sci U S A 110(1):26–33. doi:10.1073/pnas.1210514109

Schneider SH (1984) The coevolution of climate and life. Sierra Club Books, San Francisco

Self S, Widdowson M, Thordarson T, Jay AE (2006) Volatile fluxes during flood basalt eruptions and potential effects on the global environment: a Deccan perspective. Earth Planet Sci Lett 248(1–2):518–532, http://dx.doi.org/10.1016/j.epsl.2006.05.041

Sherwood SC, Nishant N (2015) Atmospheric changes through 2012 as shown by iteratively homogenized radiosonde temperature and wind data (IUKv2). Environ Res Lett 10(5):054007

Shindell DT, Faluvegi G (2009) Climate response to regional radiative forcing during the twentieth century. Nat Geosci 2(4):294–300, http://www.nature.com/ngeo/journal/v2/n4/suppinfo/ngeo473_S1.html

Shindell DT, Faluvegi G, Koch DM, Schmidt GA, Unger N, Bauer SE (2009) Improved attribution of climate forcing to emissions. Science 326(5953):716–718. doi:10.1126/science.1174760

Siegenthaler U, Oeschger H (1987) Biospheric CO_2 emissions during the past 200 years reconstructed by deconvolution of ice core data. Tellus B 39B(1–2):140–154. doi:10.1111/j.1600-0889.1987.tb00278.x

Smith SJ, Bond TC (2014) Two hundred fifty years of aerosols and climate: the end of the age of aerosols. Atmos Chem Phys 14(2):537–549. doi:10.5194/acp-14-537-2014

Smith KA, McTaggart IP, Tsuruta H (1997) Emissions of N_2O and NO associated with nitrogen fertilization in intensive agriculture, and the potential for mitigation. Soil Use Manag 13:296–304. doi:10.1111/j.1475-2743.1997.tb00601.x

Solomon S, Rosenlof KH, Portmann RW, Daniel JS, Davis SM, Sanford TJ, Plattner G-K (2010) Contributions of stratospheric water vapor to decadal changes in the rate of global warming. Science 327:1219–1223. doi:10.1126/science.1182488

Soon W, Baliunas S (2003) Proxy climatic and environmental changes of the past 1000 years. Clim Res 23(2):89–110

Sowers T (2006) Late Quaternary atmospheric CH_4 isotope record suggests marine clathrates are stable. Science 311(5762):838–840. doi:10.1126/science.1121235

Steffen W, Broadgate W, Deutsch L, Gaffney O, Ludwig C (2015) The trajectory of the Anthropocene: the great acceleration. Anthropocene Rev 2(1):81–98. doi:10.1177/2053019614564785

Streets DG, Canty T, Carmichael GR, de Foy B, Dickerson RR, Duncan BN, Edwards DP, Haynes JA, Henze DK, Houyoux MR, Jacobi DJ, Krotkov NA, Lamsal LN, Liu Y, Lu Z, Martini RV, Pfister GG, Pinder RW, Salawitch RJ, Wechti KJ (2013) Emissions estimation from satellite retrievals: a review of current capability. Atmos Environ 77:1011–1042. doi:10.1016/j.atmosenv.2013.05.051

Sturges WT, Wallington TJ, Hurley MD, Shine KP, Sihra K, Engel A, Oram DE, Penkett SA, Mulvaney R, Brenninkmeijer CAM (2000) A potent greenhouse gas identified in the atmosphere: SF_5CF_3. Science 289(5479):611–613. doi:10.1126/science.289.5479.611

Sturges WT, Oram DE, Laube JC, Reeves CE, Newland MJ, Hogan C, Martinerie P, Witrant E, Brenninkmeijer CAM, Schuck TJ, Fraser PJ (2012) Emissions halted of the potent greenhouse gas SF_5CF_3. Atmos Chem Phys 12(8):3653–3658. doi:10.5194/acp-12-3653-2012

Tans PP, Fung IY, Takahashi T (1990) Observational constraints on the global atmospheric CO_2 budget. Science 247(4949):1431–1438. doi:10.1126/science.247.4949.1431

Thomas NJ, Chang N-B, Qi C (2012) Preliminary assessment for global warming potential of leading contributory gases from a 40-in. LCD flat-screen television. Int J Life Cycle Assess 17(1):96–104. doi:10.1007/s11367-011-0341-3

Thomson AM, Calvin KV, Smith SJ, Kyle GP, Volke A, Patel P, Delgado-Arias S, Bond-Lamberty B, Wise MA, Clarke LE, Edmonds JA (2011) RCP4.5: a pathway for stabilization of radiative forcing by 2100. Clim Chang 109(1–2):77–94. doi:10.1007/s10584-011-0151-4

Tsai W-T (2008) Environmental and health risk analysis of nitrogen trifluoride (NF_3), a toxic and potent greenhouse gas. J Hazard Mater 159(2–3):257–263, http://dx.doi.org/10.1016/j.jhazmat.2008.02.023

United Nations (2015) World population prospects: the 2015 revision, methodology of the United Nations Population estimates and projections. Working Paper No. ESA/P/WP.242. New York, USA

Velders GJ, Andersen SO, Daniel JS, Fahey DW, McFarland M (2007) The importance of the Montreal Protocol in protecting climate. Proc Natl Acad Sci U S A 104(12):4814–4819. doi:10.1073/pnas.0610328104

Velders GJ, Fahey DW, Daniel JS, McFarland M, Andersen SO (2009) The large contribution of projected HFC emissions to future climate forcing. Proc Natl Acad Sci U S A 106(27):10949–10954. doi:10.1073/pnas.0902817106

Whiticar MJ (1996) Stable isotope geochemistry of coals, humic kerogens and related natural gases. Int J Coal Geol 32(1–4):191–215, http://dx.doi.org/10.1016/S0166-5162(96)00042-0

WMO (2014) World meteorological organization, scientific assessment of ozone depletion: 2014. Global Ozone Research and Monitoring Project—Report #55. Geneva, Switzerland

Yoon J, Burrows JP, Vountas M, von Hoyningen-Huene W, Chang DY, Richter A, Hilboll A (2014) Changes in atmospheric aerosol loading retrieved from space-based measurements during the past decade. Atmos Chem Phys 14(13):6881–6902. doi:10.5194/acp-14-6881-2014

Yoon J, Pozzer A, Chang DY, Lelieveld J, Kim J, Kim M, Lee YG, Koo JH, Lee J, Moon KJ (2016) Trend estimates of AERONET-observed and model-simulated AOTs between 1993 and 2013. Atmos Environ 125(Part A):33–47. http://dx.doi.org/10.1016/j.atmosenv.2015.10.058

Zeng N, Mariotti A, Wetzel P (2005) Terrestrial mechanisms of interannual CO_2 variability. Global Biogeochem Cycles 19(1). doi:10.1029/2004GB002273

Zhang Y, Wallace JM, Battisti DS (1997) ENSO-like interdecadal variability: 1900–93. J Clim 10:1004–1020

Zhang H, Wu JX, Shen ZP (2011) Radiative forcing and global warming potential of perfluorocarbons and sulfur hexafluoride. Sci China Earth Sci 54(5):764–772. doi:10.1007/s11430-010-4155-0

Zhu Z, Piao S, Myneni RB, Huang M, Zeng Z, Canadell JG, Ciais P, Sitch S, Friedlingstein P, Arneth A, Cao C, Cheng L, Kato E, Koven C, Li Y, Lian X, Liu Y, Liu R, Mao J, Pan Y, Peng S, Penuelas J, Poulter B, Pugh TAM, Stocker BD, Viovy N, Wang X, Wang Y, Xiao Z, Yang H, Zaehle S, Zeng N (2016) Greening of the Earth and its drivers. Nat Clim Change 6:791–795. doi:10.1038/nclimate3004, http://www.nature.com/nclimate/journal/vaop/ncurrent/abs/nclimate3004.html#supplementary-information

Chapter 2
Forecasting Global Warming

Austin P. Hope, Timothy P. Canty, Ross J. Salawitch,
Walter R. Tribett, and Brian F. Bennett

Abstract This chapter provides an overview of the factors that will govern the rise in global mean surface temperature (GMST) over the rest of this century. We evaluate GMST using two approaches: analysis of archived output from atmospheric, oceanic general circulation models (GCMs) and calculations conducted using a computational framework developed by our group, termed the Empirical Model of Global Climate (EM-GC). Comparison of the observed rise in GMST over the past 32 years with GCM output reveals these models tend to warm too quickly, on average by about a factor of two. Most GCMs likely represent climate feedback in a manner that amplifies the radiative forcing of climate due to greenhouse gases (GHGs) too strongly. The GCM-based forecast of GMST over the rest of the century predicts neither the target (1.5 °C) nor upper limit (2.0 °C warming) of the Paris Climate Agreement will be achieved if GHGs follow the trajectories of either the Representative Concentration Pathway (RCP) 4.5 or 8.5 scenarios. Conversely, forecasts of GMST conducted in the EM-GC framework indicate that if GHGs follow the RCP 4.5 trajectory, there is a reasonably good probability (~75 %) the Paris target of 1.5 °C warming will be achieved, and an excellent probability (>95 %) global warming will remain below 2.0 °C. Uncertainty in the EM-GC forecast of GMST is primarily caused by the ability to simulate past climate for various combinations of parameters that represent climate feedback and radiative forcing due to aerosols, which provide disparate projections of future warming.

Keywords Global warming projections • Attributable Anthropogenic Warming • Global warming hiatus • Climate feedback

2.1 Introduction

The objective of the Paris Agreement negotiated at the twenty-first session of the Conference of the Parties of the United Nations Framework Convention on Climate Change (UNFCCC) is to hold the increase in global mean surface temperature (GMST) to well below 2 °C above pre-industrial levels and to pursue efforts to limit the increase to 1.5 °C above pre-industrial levels. The rise in GMST relative to the pre-industrial baseline, termed ΔT, is the primary focus throughout this book. We consider

© The Author(s) 2017
R.J. Salawitch et al., *Paris Climate Agreement: Beacon of Hope*,
Springer Climate, DOI 10.1007/978-3-319-46939-3_2

measurements of GMST from three data centers: CRU,[1] GISS,[2] and NCEI[3] and use the latest version of each data record available at the start of summer 2016. The current values of ΔT from these data centers are 0.828 °C, 0.890 °C, and 0.848 °C respectively.[4] The rise in GMST during the past decade is more than half way to the Paris goal to limit warming to 1.5 °C. Carbon dioxide (CO_2) is the greatest waste product of modern society and global warming caused by anthropogenic release of CO_2 is on course to break through both the Paris goal and upper limit (2.0 °C) unless the world's voracious appetite for energy from the combustion of fossil fuels is soon abated.

Forecasts of ΔT are generally based on calculations conducted by general circulation models (GCMs) that have explicit representation of many processes in Earth's atmosphere and oceans. For several decades, most models have also included a treatment of the land surface and sea-ice. More recently, models have become more sophisticated by adding treatments of tropospheric aerosols, dynamic vegetation, atmospheric chemistry, and land ice. Chapter 5 of Houghton (2015) provides a good description of how GCMs operate and the evolution of these models over time.

The calculations of ΔT by GCMs considered here all use specified abundances of greenhouse gases (GHGs) and precursors of tropospheric aerosols. These specifications originate from the Representative Concentration Pathway (RCP) process that resulted in four scenarios used throughout IPCC (2013): RCP 8.5, RCP 6.0, RCP 4.5, and RCP 2.6 (van Vuuren et al. 2011a). The number following each scenario indicates the increase in radiative forcing (RF) of climate, in units of W m^{-2}, at the end of this century relative to 1750, due to the prescribed abundance of all anthropogenic GHGs. The GCMs use as input time series for the atmospheric abundance of GHGs as well as the industrial release of pollutants that are converted to aerosols. Each GCM projection of ΔT is guided by the calculation, internal to each model, of how atmospheric humidity, clouds, surface reflectivity, and ocean circulation all respond to the change in RF of climate induced by GHGs and aerosols (Houghton 2015). If the response to a specific process further increases RF of climate, it is called a ***positive feedback*** because it enhances the initial perturbation. If a response decreases RF

---

[1] The CRU temperature record is version HadCRUT4.4.0.0 from the Climatic Research Unit (CRU) of the University of East Anglia, in conjunction with the Hadley Centre of the U.K. Met Office (Jones et al. 2012), at http://www.metoffice.gov.uk/hadobs/hadcrut4/data/4.4.0.0/time_series/HadCRUT.4.4.0.0.annual_ns_avg.txt. This data record extends back to 1850.

[2] The GISS temperature record is version 3 of the Global Land-Ocean Temperature Index provided by the Goddard Institute for Space Studies (GISS) of the US National Aeronautics and Space Administration (NASA) (Hansen et al. 2010), at http://data.giss.nasa.gov/gistemp/tabledata_v3/GLB.Ts+dSST.txt. This data record extends back to 1880.

[3] The NCEI temperature record is version 3.3 of the Global Historical Climatology Network-Monthly (GHCN-M) data set provided by the National Centers for Environmental Information (NCEI) of the US National Oceanographic and Atmospheric Administration (NOAA) (Karl et al. 2015), at http://www.ncdc.noaa.gov/monitoring-references/faq/anomalies.php. This data record extends back to 1880.

[4] ΔT for CRU was found relative to the 1850–1900 baseline using data entirely from this data record; ΔT for NCEI and GISS are also for a baseline for 1850–1900, computed using a blended procedure described in the Methods note for Fig. 2.3. A decade long time period of 2006–2015 is used for this estimate of ΔT to remove the effect of year-to-year variability. A higher value of ΔT results if GMST from 2015 is used, but as explained later in this chapter, excess warmth in 2015 was due to a major El Niño Southern Oscillation event.

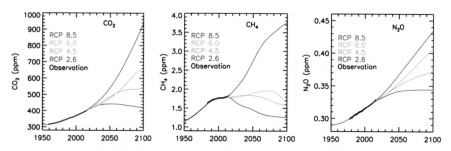

Fig. 2.1 GHG abundance, 1950–2100. Time series of the atmospheric CO_2, CH_4, and N_2O from RCP 2.6 (van Vuuren et al. 2011b), RCP 4.5 (Thomson et al. 2011), RCP 6.0 (Masui et al. 2011), RCP 8.5 (Riahi et al. 2011), and observations (*black*) (Ballantyne et al. 2012; Dlugokencky et al. 2009; Montzka et al. 2011). Values of GHG mixing ratios from RCP extend back to 1860, but this figure starts in 1950 since most of the rise in these GHGs has occurred since that time. See Methods for further information

of climate, is it called a *negative feedback*. The total effect of all responses to the prescribed perturbation to RF of climate by GHGs and aerosols is called *climate feedback*, which can vary quite a bit between GCMs, mainly due to the treatment of clouds (Bony et al. 2006; Vial et al. 2013). GCMs also provide estimates of the future evolution of precipitation, drought indices, sea-level rise, as well as variations in oceanic and atmospheric temperature and circulation (IPCC 2013).

Our focus is on analysis of projections of ΔT for the RCP 4.5 (Thomson et al. 2011) and RCP 8.5 scenarios (Riahi et al. 2011). Atmospheric abundances of the three most important anthropogenic GHGs given by the RCP 4.5 and RCP 8.5 scenarios are shown in Fig. 2.1. Under RCP 8.5, the abundances of these GHGs rise to alarmingly high levels by end of century. On the other hand, for RCP 4.5, CO_2 stabilizes at 540 parts per million by volume (ppm) (~35 % higher than contemporary level) and methane (CH_4) reaches 1.6 ppm (~10 % lower than today) in 2100. The atmospheric abundance of nitrous oxide (N_2O) continues to rise under RCP 4.5, reaching 0.37 ppm by end of century (~15 % higher than today).

The ΔRF of climate associated with RCP 4.5 and RCP 8.5 are shown in Fig. 2.2, using the grouping of GHGs defined in Chap. 1. The contrast between these two scenarios is dramatic. For RCP 4.5, ΔRF of climate levels off at mid-century, reaching 4.5 W m^{-2} at end-century. For RCP 8.5, ΔRF rises throughout the century, hitting 8.5 W m^{-2} near 2100. Both behaviors are by design (Thomson et al. 2011; Riahi et al. 2011). While CO_2 remains the most important anthropogenic GHG for both projections, other GHGs exert considerable influence.

The RCPs are meant to provide a mechanism whereby GCMs are able to simulate the response of climate for various prescribed ΔRF scenarios, in a manner that allows differences in model behavior to be assessed. Evaluation of GCM output has been greatly facilitated by the Climate Model Intercomparison Project Phase 5 (CMIP5) (Taylor et al. 2012), which maintains a computer archive of model output freely available following a simple registration procedure,[5] as well as the prior CMIP phases.

---

[5] CMIP5 GCM output is at http://cmip-pcmdi.llnl.gov/cmip5/data_getting_started.html

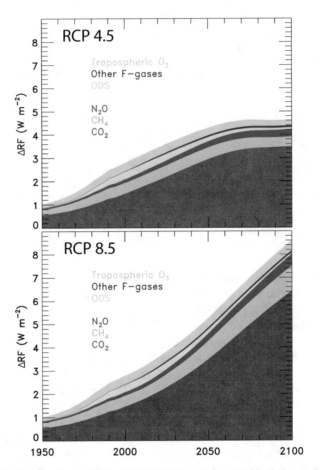

Fig. 2.2 ΔRF of climate due to GHGs, 1950–2100. Time series of ΔRF of climate, RCP 4.5 (*top*) and RCP 8.5 (*bottom*), due to the three dominant anthropogenic GHGs (CO₂, CH₄, and N₂O) plus contributions from all ozone depleting substances (ODS), other fluorine bearing compounds such as HFCs, PFCs, SF₆, and NF₃ (Other F-gases), and tropospheric O₃. *Shaded regions* represent contributions from specific gases or groups. See Methods for further information

Two other scenarios, RCP 6.0 (Masui et al. 2011) and RCP 2.6 (van Vuuren et al. 2011b), were considered by IPCC (2013). The mixing ratio of CO_2 peaks at about 670 ppm at end-century for RCP 6.0 (Fig. 2.1); the climate consequences for this scenario clearly lie between those of RCP 4.5 and RCP 8.5. For RCP 2.6, CO_2 peaks mid-century and slowly declines to 420 ppm at end-century.[6] According to the authors of RCP 2.6, this scenario "is representative of the literature on mitigation scenarios aiming to limit the increase of global mean temperature to 2 °C". While this is true for literal interpretation of the output of the GCMs that contributed to the

---

[6] Globally averaged CO_2 was ~404 ppm during summer 2016. To achieve the RCP 2.6 scenario, CO_2 at the end of the century must be comparable to the present day value.

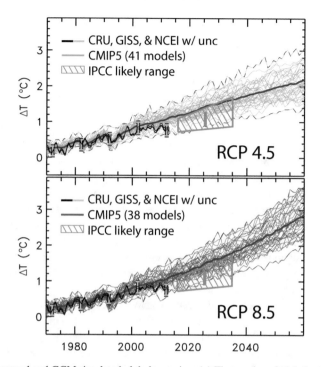

Fig. 2.3 Observed and GCM simulated global warming. (**a**) Time series of global, annually averaged ΔT relative to pre-industrial baseline from 41 GCMs that submitted output to the CMIP5 archive covering both historical and future time periods, using RCP4.5 (*light blue*). The maximum and minimum values of CMIP5 ΔT are indicated by the *dark blue dashed lines*, while the multi-model-mean is denoted by the *dark blue solid line*. Also shown are global, annually averaged observed ΔT from CRU, GISS, and NCEI (*black*) along with error bars (*grey*) that represent the uncertainty on the CRU time series. The *green trapezoid* represents the indicative likely range for annual average ΔT for 2016–2035 (i.e., top and bottom of trapezoid are upper and lower limits, respectively) and the *green bar* represents the likely range for the mean value of ΔT over 2006 to 2035, both given in Chap. 11 of IPCC (2013); (**b**) same as (**a**), expect for 38 GCMs that submitted output to the CMIP5 archive covering both historical and future time periods using RCP8.5 (*red*). After Fig. 11.25a and 11.25b of (IPCC 2013). See Methods for further information

most recent IPCC report (Rogelj et al. 2016), below we show these GCMs likely over-estimate the actual warming that will occur in the coming decades.

Figure 2.3 shows projections of ΔT from the CMIP5 GCMs found using RCP 4.5 and RCP 8.5. Observations of ΔT from CRU, NCEI, and GISS up to year 2012, as well as the CRU estimate of the uncertainty on ΔT, are shown. The green hatched trapezoid in Fig. 2.3 is the "indicative likely range for annual mean ΔT" provided by Chap. 11 of IPCC (2013).[7] Section 11.3.6.3 of this report states:

> some CMIP5 models have a higher transient response to GHGs and a larger response to other anthropogenic forcings (dominated by the effects of aerosols) than the real world (*medium confidence*). These models may warm too rapidly as GHGs increase and aerosols decline

---

[7] The trapezoid also appears in Fig. TS.14, p. 87, of the IPCC (2013) Technical Summary.

and

over the last two decades the observed rate of increase in GMST has been at the lower end
of rates simulated by CMIP5 models.

In other words, the projections of ΔT by the CMIP5 GCMs tend to be *too warm*
based on comparison of observed and modeled ΔT for prior decades (Stott et al.
2013; Gillett et al. 2013). The trapezoid shown in Fig. 2.3 represents an expert
judgement of the upper and lower limits for the evolution of ΔT over the next two
decades. The vertical bar is the likely mean value of ΔT over the 2016–2035 time
period. This projection is meant to apply to all four RCPs: i.e., it considers the full
range of possible future values for CO_2, CH_4, and N_2O between present and 2035.

Our analysis of the Paris Climate Agreement will be based on the CMIP5 GCM
output as well as calculations conducted using an Empirical Model of Global Climate
(EM-GC) developed by our group (Canty et al. 2013). The EM-GC is described in Sect.
2.2. While the EM-GC tool only calculates ΔT, this simple approach is computationally
efficient, allowing the uncertainty on ΔT of climatically important factors such as radiative forcing by tropospheric aerosols and ocean heat content to be evaluated in a rigorous
manner. We then compare estimates of how much global warming over the 1979–2010
time period can truly be attributed to human activity (Sect. 2.3). Following a brief comment on the so-called global warming hiatus (Sect. 2.4), we turn our attention to projections of ΔT (Sect. 2.5). The green trapezoid in Fig. 2.3 is featured prominently in Sect.
2.5: projections of ΔT found using the EM-GC approach are in remarkably good agreement with this IPCC (2013) expert judgement of ΔT over the next two decades, lending
credence to the accuracy of our empirically-based projections.

2.2   Empirical Model of Global Climate

Earth's climate is influenced by a variety of anthropogenic and natural factors.
Rising levels of greenhouse gases (GHGs) cause global warming (Lean and Rind
2008; Santer et al. 2013b) whereas the increased burden of tropospheric aerosols
offset a portion of the GHG-induced warming (Kiehl 2007; Smith and Bond 2014).
The most important natural drivers of climate during the past century have been the
El Niño Southern Oscillation (ENSO), the 11 year cycle in total solar irradiance
(TSI), volcanic eruptions strong enough to penetrate the tropopause as recorded by
enhanced stratospheric optical depth (SOD) (Lean and Rind 2008; Santer et al.
2013a), and variations in the strength of the Atlantic Meridional Overturning
Circulation (AMOC) (Andronova and Schlesinger 2000). Climate change is also
driven by feedbacks (changes in atmospheric water vapor, lapse rate,[8] clouds, and
the surface albedo in response to radiative forcing induced by GHGs and aerosols)

---

[8]Lapse rate is a scientific term for the variation of temperature with respect to altitude. As shown
in Fig. 1.5, over the past 50 years the upper troposphere (~10 km altitude) has warmed by a larger
amount than the surface. When this type of pattern occurs, climate scientists conclude the lapse
rate feedback is negative, because Earth's atmosphere is able to radiate heat into space more effi-
ciently. The interested reader is referred to a detailed yet accessible text entitled *Atmosphere,
Clouds, and Climate* (Randall 2012) for more information.

(Bony et al. 2006) and transport of heat from the atmosphere to the ocean that drives a long term rise in the temperature of the world's oceans (Levitus et al. 2012).

Our Empirical Model of Global Climate (EM-GC) (Canty et al. 2013) uses an approach termed multiple linear regression (MLR) to simulated observed monthly variations in the global mean surface temperature anomaly (termed ΔT_i, where i is an index representing month) using an equation that represents the various natural and anthropogenic factors that influence ΔT_i. The EM-GC formulation represents:

- RF of climate due to anthropogenic GHGs, tropospheric aerosols, and land use change
- Exchange of heat between the atmosphere and ocean, in the tropical Pacific, regulated by ENSO
- Variations in TSI reaching Earth due to the 11 year solar cycle
- Reflection of sunlight by volcanic aerosols in the stratosphere, following major eruptions
- Exchange of heat with the ocean due to variations in the strength of AMOC
- Export of heat from the atmosphere to the ocean that causes a steady long-term rise of water temperature throughout the world's oceans

The effects on ΔT of the Pacific Decadal Oscillation (PDO) (Zhang et al. 1997) and the Indian Ocean Dipole (IOD) (Saji et al. 1999) are also considered.

The hallmark of the MLR approach is that coefficients that represent the impact of GHGs, tropospheric aerosols, ENSO, major volcanoes, etc. on ΔT_i are found, such that the output of the EM-GC equations provide a good fit to the observed climate record. The most important model parameters are the total climate feedback parameter (designated λ) and a coefficient that represents the efficiency of the long-term export of heat from the atmosphere to the world's oceans (designated κ). Our approach is similar to many prior published studies, including Lean and Rind (2009), Chylek et al. (2014), Masters (2014), and Stern and Kaufmann (2014) except ocean heat export (OHE, the transfer of heat from the atmosphere to the ocean) is explicitly considered and results are presented for a wide range of model possibilities that provide reasonably good fit to the climate record, rather than relying on a single best fit. Most of the prior studies neglect OHE and typically rely on a best fit approach.

A description of the EM-GC approach is provided in the remainder of this section. While we have limited the use of equations throughout the book, they are necessary when providing a description of the model. We've concentrated the use of equations in the section that follows; comparisons of output from the EM-GC with results from the CMIP5 GCMs are presented in other sections with use of little or no equations.

2.2.1 Formulation

The Empirical Model of Global Climate (Canty et al. 2013) provides a mathematical description of observed temperature. As noted above, temperature is influenced by a variety of human and natural factors. Our approach is to compute, from the historical climate record, numerical values of the strength of climate feedback and the efficiency of the transfer of heat from the atmosphere to the ocean. We then use these two parameters to project global warming.

Here we delve into the mathematics of the EM-GC framework. Those without an appetite for the equations are encouraged to fast forward to Sect. 2.3. There will not be a quiz at the end of this chapter.

Our simulation of observed temperature involves finding values of a series of coefficients such that the model *Cost Function*:

$$Cost\ Function = \sum_{i=1}^{N_{MONTHS}} \frac{1}{\sigma_{OBSi}^2} (\Delta T_{OBSi} - \Delta T_{EM\text{-}GCi})^2 \tag{2.1}$$

is minimized. Here, $\Delta T_{OBS\ i}$ and $\Delta T_{EM\text{-}GC\ i}$ represent time series of observed and modeled monthly, global mean surface temperature anomalies, $\sigma_{OBS\ i}$ is the 1-sigma uncertainty associated with each temperature observation, i is an index for month, and N_{MONTHS} is the total number of months. The use of $\sigma_{OBS\ i}^2$ in the denominator of Eq. 2.1 forces modeled $\Delta T_{EM\text{-}GC\ i}$ to lie closest to data with smaller uncertainty, which tends to be the latter half of the $\Delta T_{OBS\ i}$ record.

The expression for $\Delta T_{EM\text{-}GC\ i}$ is:

$$\begin{aligned}
\Delta T_{EM\text{-}GCi} = \frac{1+\gamma}{\lambda_P} & \{(GHG\ \Delta RF_i + Aerosol\ \Delta RF_i + LUC\ \Delta RF_i\} + C_0 \\
& + C_1 \times SOD_{i-6} + C_2 \times TSI_{i-1} + C_3 \times ENSO_{i-3} \\
& + C_4 \times AMV_i + C_5 \times PDO_i + C_6 \times IOD_i \\
& - \frac{Q_{OCEANi}}{\lambda_P}
\end{aligned} \tag{2.2}$$

where model input variables (described immediately below) are used to calculate the model output parameters C_i and γ. In Eq. 2.2 GHG ΔRF_i, Aerosol ΔRF_i, and LUC ΔRF_i represent monthly time series of the ΔRF of climate due to anthropogenic GHGs, tropospheric aerosol, and land use change; $\lambda_P = 3.2$ W m^{-2} °C^{-1} is the response of surface temperature to a RF perturbation in the absence of climate feedback ("P" is used as a subscript because this term is called the Planck response function by the climate modeling community (Bony et al. 2006)); SOD_{i-6}, TSI_{i-1}, $ENSO_{i-3}$ represent indices for stratospheric optical depth, total solar irradiance, and El Niño Southern Oscillation lagged by 6 months, 1 month, and 3 months, respectively; AMV_i, PDO_i, and IOD_i represent indices for Atlantic Multidecadal Variability (a proxy for the strength of AMOC), the Pacific Decadal Oscillation, and the Indian Ocean Dipole; and $Q_{OCEAN\ i} / \lambda_P$ is the Ocean Heat Export term. The use of temporal lags for SOD, TSI, and ENSO is common for MLR approaches: Lean and Rind (2008) use lags of 6 months, 1 month and 4 months, respectively, for these terms. These lags represent the delay between forcing of the climate system and the response of RF of climate at the tropopause, after stratospheric adjustment. These lags are discussed at length in our model description paper (Canty et al. 2013). Finally, the AMV, PDO, and IOD terms have traditionally not been used in MLR models. Below, results are shown with and without consideration of these three terms. No lag is imposed for these three terms since the indices used to describe these processes vary slowly with respect to time.

The coefficients (C_1 to C_6) that multiply the various model terms, as well as the constant term C_0 and the variable γ, are found using multiple linear regression, which provides numerical values for each of these parameters such that the *Cost Function* (Eq. 2.1) has the smallest possible value. The term γ in Eq. 2.2 is the dimensionless climate sensitivity parameter. If the net response of changes in humidity, lapse rate, clouds, and surface albedo that occur in response to anthropogenic ΔRF of climate is positive, as is most often the case, then the value of γ is positive.

The estimate of Q_{OCEAN} is based on finding the value of the final model output parameter κ, the ocean heat uptake efficiency coefficient with units of W m^{-2} °C^{-1} (Raper et al. 2002) that best fits a time series of ocean heat content (OHC), where:

$$Q_{OCEANi} = \kappa \frac{1+\gamma}{\lambda_P}(GHG\, \Delta RF_{i-72} + Aerosol\, \Delta RF_{i-72} + LUC\, \Delta RF_{i-72}) \quad (2.3)$$

The subscripts $i - 72$ in Eq. 2.3 represent a 6 year (or 72 month) lag between the anthropogenic ΔRF perturbation and the export of heat to the upper ocean. The numerical estimate of this lag is based on the simulations described by Schwartz (2012); the projections of global warming found using the EM-GC framework are insensitive to any reasonable choice for the this lag. Since the model is based on matching perturbations in RF of climate to variations in temperature, the flow of heat from the atmosphere to the ocean is modeled as a perturbation to the mean state induced by anthropogenic RF of climate (i.e., Q_{OCEAN} in Eq. 2.2 depends only on "delta" terms that represent human influence on climate). Finally, the net effect of human activity on ΔT is the sum of GHG warming, aerosol cooling, very slight cooling due to land use change, and ocean heat export:

$$\Delta T^{HUMAN}_{i} = \frac{1}{\lambda_P}[(1+\gamma)(GHG\, \Delta RF_i + Aerosol\, \Delta RF_i + LUC\, \Delta RF_i) - Q_{OCEANi}] \quad (2.4)$$

Equations 2.1–2.4 constitute our Empirical Model of Global Climate. Of the model inputs, the aerosol ΔRF term is the most uncertain. As shown below, there is a strong relation between the value of the climate sensitivity parameter γ and the magnitude of aerosol ΔRF. This dependency is well known in the climate community, as discussed for example by Kiehl (2007). Also, there is a wide variation in the value of κ, depending on which dataset is used to specify OHC.

Figures 2.4 and 2.5 provide a graphical illustration of how the model works. The simulations in these figures use estimates for GHG and aerosol ΔRF from RCP 4.5, tied to the best estimate for aerosol ΔRF in year 2011 (AerRF$_{2011}$) of −0.9 W m^{-2} from IPCC (2013), and a time series for OHC in the upper 700 m of the global oceans that is an average of six published studies. In the interest of keeping the attention of those reading this far, we describe a few simulations prior to delving into further details about the model parameters.

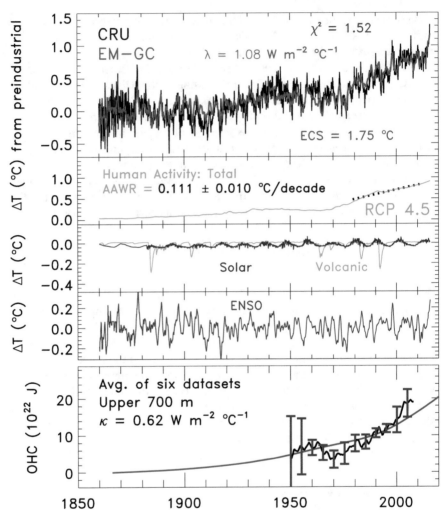

Fig. 2.4 Observed and EM-GC simulated global warming, 1860–2015. Ladder plot showing CRU observed global, monthly mean ΔT from CRU (*black*) and as simulated by the EM-GC (*red*), both relative to pre-industrial baseline (*top rung*); the contribution to ΔT from humans (*orange*) (*second rung*), and contributions from natural sources of climate variability due to fluctuations in the output of the sun and major volcanic eruptions (*third rung*), and ENSO (*fourth rung*). The final rung compares modeled and measured ocean heat content (OHC), where the data show the average (used in the model) and standard deviation of OHC from six data sets. See Methods for further information

Figure 2.4 is a so-called "ladder plot" that compares a time series of observed, monthly values of ΔT (top rung) from CRU (black) to the output of the model (red). For the simulation in Fig. 2.4, the AMV, PDO, and IOD terms have been neglected. The model provides a reasonably good description of the observed global temperature anomaly. The red curve on the top panel is the sum of the orange curve on the second panel (total effect of human activity), the blue and purple curves on the third panel (volcanic and solar terms), and the cardinal curve on the fourth panel (ENSO), plus the

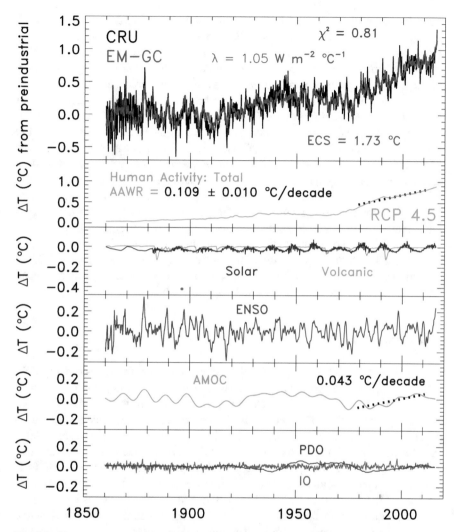

Fig. 2.5 Observed and EM-GC simulated global warming, 1860–2015. Same as Fig. 2.4, except the EM-GC equations have been expanded to include the effects of the Atlantic Meridional Overturning Circulation (AMOC), the Pacific Decadal Oscillation (PDO), and the Indian Ocean Dipole (IOD). The fifth rung of the ladder plot shows contributions to variations in ΔT from fluctuations in the strength of the AMOC; the sixth rung shows contributions from PDO and IOD. See Methods for further information

regression constant C_0 (not shown). Finally, the bottom panel shows a comparison of a time series of OHC (available only from 1950 to 2007) to the modeled Q_{OCEAN} term.

Figure 2.5 is similar to Fig. 2.4, except here the model has been expanded to include the AMV, PDO, and IOD terms in Eq. 2.2. The OHC comparison is not shown in Fig. 2.5 because it looks identical to the bottom panel of Fig. 2.4. The red curve on the top panel of Fig. 2.5 is the sum of the curves shown in the rest of the panels, plus the constant C_0. The top panel of Fig. 2.5 shows remarkably good agreement between observed ΔT from CRU (black) and modeled ΔT found using the EM-GC equation (red).

Consideration of these three additional ocean proxies improves the simulation of ΔT around year 1910 and in the mid-1940s (Fig. 2.5) compared to the results shown in Fig. 2.4, which lacked these terms. Most of this improvement is due to the use of AMV as a proxy for variations in the strength of the Atlantic Meridional Overturning Circulation, which only recently has been recognized as having a considerable effect on global climate (Schlesinger and Ramankutty 1994; Andronova and Schlesinger 2000). In our approach, the PDO (Zhang et al. 1997) and the IOD (Saji et al. 1999) have little expression on global climate, which is a common finding using MLR analysis of the ~150 year long record of ΔT (Rypdal 2015; Chylek et al. 2014). Also, upon inclusion of the AMV proxy (Fig. 2.5), the cooling after major volcanic eruptions is diminished by nearly a factor of two relative to a MLR analysis that neglects this term (volcanic term in Fig. 2.5 compared to volcanic term in Fig. 2.4). This finding could have significant implications for the use of volcanic cooling as a proxy for the efficacy of geo-engineering of climate via stratospheric sulfate injection (Canty et al. 2013).

Additional detail on inputs to the Empirical Model of Global Climate is provided in Sect. 2.2.1.1. More explanation of the model outputs is given in Sect. 2.2.1.2. Both of these sections are condensed from our model description paper (Canty et al. 2013), including a few updates since the original publication.

2.2.1.1 Model Inputs

The ΔRF due to GHGs is based on global, annual mean mixing ratios of CO_2, CH_4, N_2O, the class of halogenated compounds known as ozone depleting substances (ODS), HFCs, PFCs, SF_6, and NF_3 (Other F-gases) provided by the RCP 4.5 (Thomson et al. 2011) and RCP 8.5 (Riahi et al. 2011) scenarios. Annual abundances are interpolated to a monthly time grid, because monthly resolution is needed to resolve short-term impacts on ΔT of processes such as ENSO and volcanic eruptions. Values of ΔRF for each GHG are computed using formula originally given in Table 6.2 of IPCC (2001) except the pre-industrial value of CH_4 has been adjusted to 0.722 ppm, following Table AII.1.1a of (IPCC 2013). The ΔRF due to tropospheric O_3 is based on the work of Meinshausen et al. (2011), obtained from a file posted at the Potsdam Institute for Climate Impact Research website. The sum of ΔRF due to CO_2, CH_4, N_2O, ODS, Other F-gases, and tropospheric O_3 constitutes GHG ΔRF_i in Eq. 2.2.

The ΔRF due to aerosols is the sum of direct and indirect effects of six types of aerosols, as described in Sect. 3.2.2 of Canty et al. (2013). The six aerosol types are sulfate, mineral dust, ammonium nitrate, fossil fuel organic carbon, fossil fuel black carbon, and biomass burning emissions of organic and black carbon. The direct ΔRF for all aerosol types other than sulfate is also based on the work of Meinshausen et al. (2011), again obtained from files posted at the Potsdam Institute for Climate Impact Research website. Different estimates for RCP 4.5 and RCP 8.5 are used, since it is assumed that reduction of atmospheric release of aerosol precursors will occur more quickly in RCP 4.5, in lock-step with the decreased emission of GHGs in this scenario relative to RCP 8.5. The direct RF due to sulfate is based on the work of Smith et al. (2011). Scaling parameters are used to multiply the direct ΔRF of aerosols, to account for the aerosol indirect effect, as described in Sect. 3.2.2 of Canty et al. (2013).

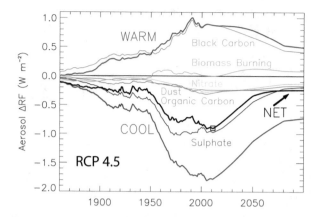

Fig. 2.6 Aerosol ΔRF versus time, RCP 4.5, for AerRF$_{2011}$ = −0.9 W m^{-2} (*open square*), The figure shows ΔRF for six aerosol components (as indicated), the sum ΔRF for all aerosols that warm (*red*), the sum of ΔRF for all aerosols that cool (*blue*), and the net ΔRF of aerosols (*black*). See Methods for further information

Figure 2.6 shows total ΔRF (black line) due to tropospheric aerosols that was used as EM-GC input (i.e., the term Aerosol ΔRF$_i$ in Eq. 2.2) for the calculations shown in Figs. 2.4 and 2.5, as well as the contribution to aerosol ΔRF from the six classes of aerosols. This particular time series, based on RCP 4.5, has been designed to match the IPCC (2013) best estimate of AerRF$_{2011}$ (aerosol ΔRF in year 2011) of −0.9 W m^{-2}.

As detailed in Canty et al. (2013), a specific value of AerRF$_{2011}$ can be found using a variety of combinations of scaling parameters that account for the aerosol indirect effect. Figure 2.7a shows time series of aerosol ΔRF for RCP 4.5 designed to match five rather disparate estimates of AerRF$_{2011}$ from IPCC (2013):

- −0.9 W m^{-2} (*best estimate*)
- −0.4 and −1.5 W m^{-2} (upper and lower limits of the *likely range*, denoted by the upper and lower edges of rectangle marked "Expert Judgement" in Fig. 7.19b of IPCC (2013), which are the 17th and 83^d percentiles of the estimated distribution)
- −0.1 and −1.9 W m^{-2} (upper and lower limits of the *possible range*, denoted by the error bars on the "Expert Judgement" rectangle in Fig. 7.19b, which are the 5th and 95th percentiles of the estimated distribution)

Figure 2.7b shows aerosol ΔRF designed to match these same five values of AerRF$_{2011}$, except for the RCP 8.5 emission of aerosol precursors. Three estimates of Aerosol ΔRF are shown for each value of AerRF$_{2011}$, found using scaling parameters described in Methods.

Variations in the RF of climate due to the land use change (LUC) is the final anthropogenic term considered in our EM-GC. Numerical values of LUC ΔRF$_i$ in Eq. 2.2 are based on Table AII.1.2 of IPCC (2013). This term, which has an extremely minor effect on computed ΔT and is included for completeness, represents changes in the reflectivity of Earth's surface caused, for example, by conversion of forest to concrete. The release of carbon and other GHGs due to LUC is not represented by this term, but rather by the GHG ΔRF$_i$ term.

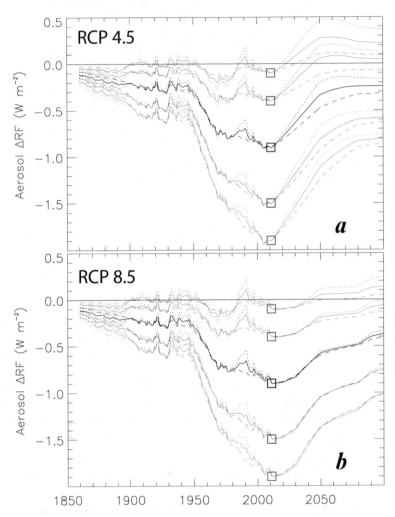

Fig. 2.7 Aerosol ΔRF versus time, RCP 4.5 and 8.5. (**a**) Various scenarios for $AerRF_{2011}$ of -0.1. -0.4, -0.9, -1.5, and -1.9 W m^{-2} (*open squares*) for RCP 4.5 aerosol precursor emissions; (**b**) same as (**a**), except for RCP 8.5 emission scenarios. See Methods for further information

We next describe data used to define EM-GC inputs of stratospheric optical depth (SOD), total solar irradiance (TSI), El Niño Southern Oscillation (ENSO), Atlantic Multidecadal Variability (AMV), Pacific Decadal Oscillation (PDO), and the Indian Ocean Dipole (IOD). These measurements are discussed in considerable detail by Canty et al. (2013); therefore, only brief descriptions are given here.

The time series for SOD_i in Eq. 2.2 is based on the global, monthly mean data set of Sato et al. (1993), available from 1850 to the end of 2012.[9] This time series makes use of ground-based, balloon-borne, and satellite observations, and represents pertur-

---

[9]The Sato et al. (1993) SOD record is at: http://data.giss.nasa.gov/modelforce/strataer/tau. line_2012.12.txt

bations to the stratospheric sulfate aerosol layer induced by volcanic eruptions that are energetic enough to penetrate the tropopause. The Sato et al. (1993) dataset compares reasonably well with an independent estimate of SOD provided by Ammann et al. (2003), which is based on a four-member ensemble simulation of volcanic eruptions by a GCM that resolves the troposphere and stratosphere and is available from 1890 to 2008 (Fig. 2.18 of IPCC IPCC 2007). The value of SOD is held constant at 0.0035 for October 2012 onwards, due to unavailability of data from the Sato et al. (1993) for more recent periods of time. The Sato et al. (1993) SOD record resolves the recent eruptions of Kasatochi, Sarychev and Nabro (Rieger et al. 2015; Fromm et al. 2014), but stops short of the April 2015 eruption of Calbuco that deposited sulfate into the high latitude, summer stratosphere (Solomon et al. 2016). Since the perturbation to global SOD due to volcanic eruptions between the end of 2012 and summer 2016 is small, the use of a constant value for SOD since October 2012 has no bearing on any of our scientific conclusions. The use of $i - 6$ as the subscript for SOD in Eq. 2.2 represents a 6 month delay between volcanic forcing and surface temperature response; a delay of ~6 months was found by the thermodynamic analyses of Douglass and Knox (2005) and Thompson et al. (2009) and a 6 month delay is used in the MLR studies of Lean and Rind (2008) and Foster and Rahmstorf (2011).

The time series of TSI_i in Eq. 2.2 is based on two data sets. For years prior to 1978, TSI originates from reconstructions that make use of the number, location, and darkening of sunspots as well as various measurements from ground-based solar observatories (Lean 2000; Wang et al. 2005). Since 1978, TSI is based on various-spaced based measurements. The magnitude of TSI varies with the well characterized 11 year sunspot cycle, due to distortion of magnetic field lines caused by differential rotation of the sun.[10] A 1 month lag for TSI_i is used in Eq. 2.2 because this yields the largest value of C_2, the common approach for defining slight temporal offset between perturbation (solar output) and response (global temperature) in MLR-based models (Lean and Rind 2008).

The time series of $ENSO_i$ in Eq. 2.2 is based on the Tropical Pacific Index (TPI), computed as described by Zhang et al. (1997). This index represents the anomaly of sea surface temperature (SST) in the region bounded by 20°S to 20°N latitude and 160°E to 80°W longitude, relative to a long-term climatology. The SST record of HadSST3.1.1.0 (Kennedy et al. 2011a, b)[11] has been used to compute TPI. A 3 month lag has been applied to ENSO, because this provides the highest correlation between TPI and a simulated response of GMST to ENSO that was computed using a thermodynamic approach (Thompson et al. 2009).

The time series for AMV_i in Eq. 2.2 is based on the time evolution of area weighted, monthly mean SST in the Atlantic Ocean, between the equator and 60°N (Schlesinger

---

[10] TSI for start of 2009–2015 is from column 3 of: ftp://ftp.pmodwrc.ch/pub/data/irradiance/composite/DataPlots/composite_*.dat where * is used because the name of this file changes as it is regularly updated.

TSI from 1882 to end of 2008 is from column 3 of : https://ftp.geomar.de/users/kmatthes/CMIP5 TSI prior to 1882 is from column 2 of: ftp://ftp.ncdc.noaa.gov/pub/data/paleo/climate_forcing/solar_variability/lean2000_irradiance.txt

[11] HadSST3.1.1.0 data are at: http://hadobs.metoffice.com/hadsst3/data/HadSST.3.1.1.0/netcdf/HadSST.3.1.1.0.median_netcdf.zip

and Ramankutty 1994). Here, data from HadSST3.1.1.0 have been used (same citations and web address as for ENSO). As shown in the Supplement of Canty et al. (2013), nearly identical scientific results are obtained using SST from NOAA. The AMV index is a proxy for changes in the strength of the Atlantic Meridional Overturning Circulation (AMOC) (Knight et al. 2005; Stouffer et al. 2006; Zhang et al. 2007; Medhaug and Furevik 2011). Others use Atlantic Multidecadal Oscillation (AMO) to describe this index, but we prefer AMV because whether or not the strength of the AMOC varies in a purely oscillatory manner (Vincze and Jánosi 2011) is of no consequence to the use of this proxy in the EM-GC framework.

There are two important details regarding AMV_i that bear mentioning. This index represents the fact that, during times of increased strength of the AMOC, the ocean releases more heat to the atmosphere.[12] There is considerable debate regarding whether the strength of AMOC varies over time (e.g., Box 5.1 of IPCC (2007) and Willis (2010)). Our focus is on anomalies of AMOC over time; hence, the AMV_i index is de-trended.[13] As shown in Fig. 5 of Canty et al. (2013), various choices for how this index is de-trended have considerable effect on the shape of the resulting time series, which is important for the EM-GC approach. Here, total anthropogenic ΔRF of climate is used to de-trend AMV_i, because this method appears to provide a more realistic means to infer variations in the strength of AMOC from the North Atlantic SST record than other de-trending options (Canty et al. 2013). The second detail involves whether monthly data should be used for the AMV_i index, since the AMOC is sluggish and variations of North Atlantic SST on time scales of a year or less likely do not represent variations in large-scale, ocean circulation. Throughout this chapter, the AMV_i index has been filtered to remove all components with temporal variations shorter than 9 years; only variations of SST on time scales of a decade or longer are preserved. The interested reader is invited to examine Fig. 7 of (Canty et al. 2013) to see the impact of various options for how AMV_i is filtered.

A major international research effort has provided new insight into temporal variations of the strength of AMOC (Srokosz and Bryden 2015). The RAPID-AMOC program, led by the Natural Environment Research Council of the United Kingdom, is designed to monitor the strength of the AMOC by deployment of an array of instruments at 26.5°N latitude, across the Atlantic Ocean, which measure temperature, salinity and ocean water velocities from the surface to ocean floor (Duchez et al. 2014). Analysis of a 10 year (2004–2014) time series of data reveals a decline in the strength of AMOC over this decade, similar to that shown by our proxy (AMOC ladder, Fig. 2.5) over this same period of time.

---

[12] An illustration of the physics of the interplay between AMOC and release of heat to the atmosphere from the ocean is at http://www.whoi.edu/cms/images/oceanus/2006/11/nao-en_33957.jpg

[13] The de-trending of AMV, the proxy for variations in the strength of AMOC, means that when examined over the *entire* 156 year record of the simulation, the slope of the panel marked AMOC in Fig 2.5 is near zero. The proxy used to represent AMOC is based on measurements of sea surface temperature, which rise over time due to the transfer of heat from the atmosphere to the ocean. Within an MLR model such as the EM-GC, the AMOC proxy should be de-trended, or else a number of erroneous conclusions regarding long-term climate change could result. See Sect. 3.2.3 of Canty et al. (2013) for further discussion.

The PDO represents the temporal evolution of specific patterns of sea level pressure and temperature of the Pacific Ocean poleward of 20°N (Zhang et al. 1997), which is caused by the response of the ocean to spatially coherent atmospheric forcing (Saravanan and McWilliams 1998; Wu and Liu 2003). The PDO is of considerable interest because variations correlate with the productivity of the fishing industry in the Pacific (Chavez et al. 2003). An index based on analysis of the patterns of SST conducted by the University of Washington[14] is used.

The IOD index[15] represents the temperature gradient between the Western and Southeastern portions of the equatorial Indian Ocean (Saji et al. 1999). The IOD index is used so that all three major ocean basins are represented. Variations in the IOD have important regional effects, including rainfall in Australia (Cai et al. 2011). However, global effects are small, most likely due to the small size of the Indian Ocean relative to the Atlantic and Pacific oceans.

The increase in the RF of climate due to human activity causes a rise in temperature of both the atmosphere and the water column of the world's oceans (Raper et al. 2002; Hansen et al. 2011; Schwartz 2012). The oceanographic community has used measurements of temperature throughout the water column, obtained by a variety of sensor systems and data assimilation techniques, to estimate the time variation of the heat content of the world's oceans (OHC, or Ocean Heat Content) (Carton and Santorelli 2008). Generally the focus has been on the upper 700 m of the oceans.

Considerable uncertainty exists in OHC. Figure 2.8 shows estimates of OHC in the upper 700 m of the world's oceans from six studies: Ishii and Kimoto (2009), Carton and Giese (2008), Balmaseda et al. (2013), Levitus et al. (2012), Church et al. (2011), Gouretski and Reseghetti (2010) as well as the average of the data from these six studies. Ostensibly, all of the studies make use of similar (if not the same) measurements from expendable bathy-thermograph (XBT) devices and the more accurate conductivity temperature depth (CTD) probes. Use of CTDs began in the 1980s, and expanded considerably in 2001 based on the deployment of thousands of drifting floats under the Argo program (Riser et al. 2016). Alas, the ocean is vast and much is not sampled. The differences in OHC shown in Fig. 2.8 published by various groups represent different methods to fill in regions not sampled by CTDs, as well as various assumptions regarding the calibration (including fall rate correction) of data returned by XBTs.

The $Q_{OCEAN\,i}$ term in Eq. 2.3 is the EM-GC representation of OHE in units of W m^{-2}: i.e., OHE is heat flux. The quantity OHC represents the energy content of the upper 700 m of the world's oceans. To relate OHC and OHE, several computational steps are necessary. First, the OHC values shown in Fig. 2.8 are multiplied by 1.42 (which equals 1/0.7) to account for the estimate that 70 % of the rise in OHC of the

---

[14] The PDO index is at http://research.jisao.washington.edu/pdo/PDO. This record begins in year 1900. Prior to 1900 we assume PDO$_i$ is equal to 0.

[15] The index for IOD from 1982 to present is based on this record provided by the Observing System Monitoring Center of NOAA http://stateoftheocean.osmc.noaa.gov/sur/data/dmi.nc

From 1860 to 1981, IOD is based on data provided by the Japan Agency for Marine-Earth Science and Technology at http://www.jamstec.go.jp/frcgc/research/d1/iod/kaplan_sst_dmi_new.txt

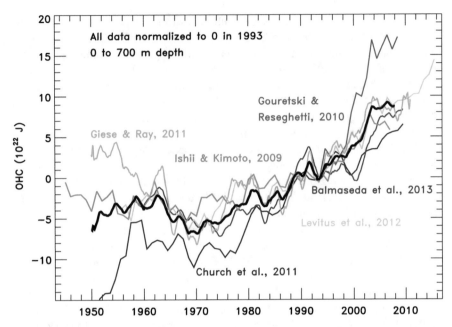

Fig. 2.8 Ocean Heat Content (OHC) versus time from six sources (*colored*, as indicated). The *black solid line* is the average of the six measurements used in most of the EM-GC calculations. See Methods for further information

world's oceans occurs in the upper 700 m (Sect. 5.2.2.1 of IPCC 2007). This multiplication is carried out because ocean heat export in the model must represent the entire water column. As stated above, a 6 year lag is assumed between perturbation and response (Schwartz 2012). Next, OHC is divided by 3.3×10^{14} m², the surface area of the world's oceans. Finally, a value for κ is derived so that the change in OHC over the period of time covered by a particular data set (i.e., the average time derivative) is matched, rather than attempting to model the ups and downs of any particular OHC record. Since the ups and downs of the various records are uncorrelated, it is more likely these variations reflect measurement noise rather than true signal.

2.2.1.2 Model Outputs

In addition to the regression coefficients, two additional parameters are found by the EM-GC: the climate sensitivity parameter (γ in Eq. 2.2) and the ocean heat uptake efficiency coefficient (κ in Eq. 2.3). As described in Sect. 2.5, values of γ and κ inferred from the prior climate record are used to obtain projections of ΔT, assuming γ and κ remain constant in time. In this section, some context for the numerical values of γ and κ is presented. Two additional model output terms, the climate feedback parameter (λ) and Equilibrium Climate Sensitivity (ECS), both of which are found from γ, are described. Finally, a metric for model performance, χ^2, which plays an important role for the projections of ΔT, is defined.

The value of κ found using the OHC record for the upper 700 m of the world's oceans, averaged from six studies, is 0.62 W m^{-2} °C^{-1} (bottom panel, Fig. 2.4). As stated in Sect. 2.2.1.1, the calculation of κ considers the increase in temperature for depths below 700 m by scaling observations from the upper part of the ocean. Of the six OHC datasets, Ishii and Kimoto (2009) results in the smallest value for κ (0.43 W m^{-2} °C^{-1}) and Gouretski and Reseghetti (2010) leads to the largest value (1.52 W m^{-2} °C^{-1}). All of the values of κ found using various time series for OHC fall within the range of empirical estimates and coupled ocean-atmosphere model behavior that is shown in Fig. 2 of Raper et al. (2002). As such, the representation of ocean heat export in the EM-GC framework is realistic, given the present state of knowledge. If the true value of κ changes over time, then our projections of ΔT based on an assumption of constant κ will require modification. Past measurements of OHC are too uncertain to infer, from the prior record, whether κ has changed. The nearly factor of 3 difference in κ inferred from various, credible estimates of OHC is certainly much larger than any reasonable change in κ that could have occurred during the time of OHC observations.

The value of γ found for the EM-GC simulation shown in Fig. 2.5 is 0.49. This means the increase in RF of climate due to GHGs, tropospheric aerosols, and land use change from 1860 to present must be increased by ~50 % (i.e., multiplied by 1.49) to obtain best fit to observed ΔT. In other words, the sum of all climate feedbacks must be positive. Model parameter γ represents the sensitivity of climate to all of the feedbacks that occur in response to the perturbation to RF at the tropopause induced by humans, and is related to the climate feedback parameter λ via:

$$1 + \gamma = \frac{1}{1 - \dfrac{\lambda}{\lambda_P}}$$

where $\lambda = \sum$ All Climate Feedbacks   (2.5)

i.e. $\lambda = \lambda_{WATER VAPOR} + \lambda_{CLOUDS} + \lambda_{LAPSE RATE} + \lambda_{SURFACE REFLECTIVITY} + $ etc.

This formulation for the relation between γ and λ is commonly used in the climate modeling community (see Sect. 8.6 of IPCC (2007)). We record λ rather than γ on all of the EM-GC ladder plots (Figs. 2.4 and 2.5) because λ is more directly comparable to GCM output, such as that in Table 9.5 of IPCC (2013).

Equilibrium climate sensitivity (ECS) is also given on the top rung of the EM-GC ladder plots. This metric represents the increase in ΔT of the climate system after it has attained equilibrium, in response to a doubling of atmospheric CO_2. In the EM-GC framework ECS is expressed as[16]:

$$ECS = \frac{1+\gamma}{\lambda_P} 3.71\,W\,m^{-2} \qquad (2.6)$$

ECS is often used to compare and evaluate climate simulations. The EM-GC run shown in Fig. 2.5 has an ECS of 1.73 °C, which means that if CO_2 were to double (i.e., reach 560 ppm, twice the pre-industrial value of 280 ppm) and if all other GHGs were to remain constant at their pre-industrial level, then ΔT would rise to a level about midway between the Paris target (1.5 °C) and upper limit (2.0 °C). As will soon be shown, ECS is a difficult metric to use for evaluating climate models because it depends rather sensitively on both aerosol ΔRF and ocean heat content, both of which have considerable uncertainty.

The top rung of each EM-GC ladder plot also contains a numerical value for reduced chi-squared (χ^2), a parameter that defines the goodness of fit between a series of observed and modeled quantities. In our framework, χ^2 is defined as:

$$\chi^2 = \frac{1}{(N_{YEARS} - N_{FITTING\ PARAMETERS} - 1)} \times \sum_{j=1}^{N_{YEARS}} \frac{1}{(\sigma_{OBS\,j}^{\,2})} \left(\langle \Delta T_{OBS\,j} \rangle - \langle \Delta T_{EM\text{-}GC\,j} \rangle \right)^2 \qquad (2.7)$$

where $\langle \Delta T_{OBS\,j} \rangle$, $\langle \Delta T_{EM\text{-}GC\,j} \rangle$, and $\langle \sigma_{OBS\,j} \rangle$, represent the annually averaged observed temperature anomaly, the annually averaged modeled temperature anomaly, and the uncertainty of the annually averaged observed temperature anomaly, respectively, and $N_{FITTING\ PAREMETERS}$ equals 6 for the simulation shown in Fig. 2.4 (four regression coefficients plus the two parameters γ and κ) and equals 9 for Fig. 2.5 (three additional regression coefficients). The formula for χ^2 is expressed in terms of annual averages, rather than monthly values, due to the statistical behavior of the two time series that appear in the formula.[17]

---

[16] The derivation is:

$$ECS = \frac{1+\gamma}{\lambda_P} \Delta RF_{CO2} = \frac{1+\gamma}{\lambda_P} 5.35\,W\,m^{-2}\,\ln\frac{CO_2^{FINAL}}{CO_2^{INITIAL}} = \frac{1+\gamma}{\lambda_P} 5.35\,W\,m^{-2}\,\ln(2) = \frac{1+\gamma}{\lambda_P} 3.71\,W\,m^{-2}$$

if we assume a doubling of atmospheric CO_2.

The expression for ΔRF_{CO2} is from Myhre et al. (1998).

[17] For those familiar with statistics, the auto-correlation function of modeled ΔT is compared to the auto-correlation function of the measured ΔT. As shown in the supplement to Canty et al. (2013), these functions differ considerably for comparison of measured and modeled monthly anomalies, indicating either the presence of a forcing in the system not resolved by the model or else considerable noise in the measurement. These auto-correlation functions are quite similar for comparison of measured and modeled annual anomalies, indicating proper physical structure of the modeled quantity and *appropriate use* of χ^2, if applied to *annual averages* of both modeled and measured anomalies.

The EM-GC simulation in Fig. 2.4 has $\chi^2 = 1.52$. In the world of physics, this would be termed a reasonably good model simulation. Such an impression is also apparent based on visual inspection of the red and black curves on the top rung of Fig. 2.4. The EM-GC simulation in Fig. 2.5 has $\chi^2 = 0.81$, which is an exceptionally good simulation both in the literal interpretation of χ^2, as well as visual inspection of Fig. 2.5. For the quantitative assessments of the amount of global warming that can be attributed to humans, as well as the projections of future global warming, EM-GC simulations are weighted by $1/\chi^2$, such that the better the goodness of fit (i.e., the smaller the value of χ^2) the larger the weight. Chapter 7 of Taylor (1982) provides a description of the utility of this weighting approach.

2.2.1.3 The Degeneracy of Earth's Climate

Figure 2.9 shows simulations of Earth's climate that differ only due to choice of ΔRF due to tropospheric aerosols. Figure 2.9a shows results for $AerRF_{2011}$ of -0.4 W m^{-2} (upper limit of IPCC (2013) likely range), -0.9 W m^{-2} (IPCC best estimate), and -1.5 W m^{-2} (lower limit of IPCC likely range). For each simulation, the upper rung of a typical EM-GC ladder plot is shown, but with ΔT projected into the future. Projections use values of λ and κ associated with each simulation, together with RCP 4.5 for GHG abundances and aerosol precursor emissions. Each simulation uses the OHC record based on the average of the six studies shown in Fig. 2.8. For our projections of ΔT, the only term considered is ΔT^{HUMAN} (Eq. 2.4): i.e., we assume that the future change in temperature will be based on GHG warming and aerosol cooling from RCP 4.5, climate feedback, and ocean heat export. It is also assumed that natural factors such as ENSO, solar, and volcanoes will have no influence on future temperature. The second rung of Fig. 2.9 shows ΔT^{HUMAN} as well as the contributions from individual terms (here the OHE term is not shown for clarity because it is small and nearly the same for each simulation[18]). The GMST experienced in 2015 was unusually large due to the effect of ENSO, which is illustrated by inclusion of the ENSO rung for Fig. 2.9b.[19]

Figure 2.9 shows that the climate record can be fit nearly equally well using the EM-GC approach for two contrasting scenarios:

(1) tropospheric aerosols have had little overall effect on prior climate due to a near balance of cooling (primarily sulfate aerosols) and heating (primarily black carbon aerosols) and the **climate feedback** (numerical value of λ) needed to fit observed ΔT_i is **small** (Fig. 2.9a).

---

[18] Time series of ocean heat export (OHE) appear on the next figure, which illustrates the sensitivity of the EM-GC model to choice of data set for ocean heat content (OHC).

[19] The ENSO rungs for Fig. 2.9a, c are nearly identical to Fig 2.9b and is only shown once

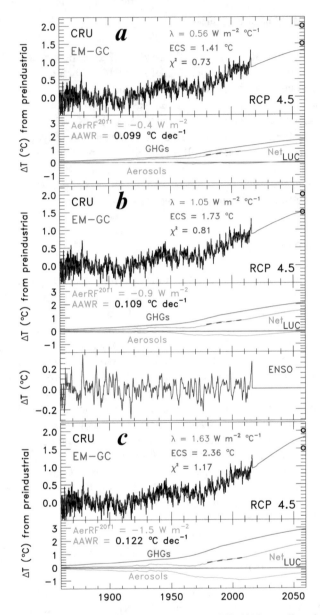

Fig. 2.9 Observed and EM-GC simulated global warming, 1860–2015 as well as global warming projected to 2060. (**a**) Top rung of a typical ladder plot, comparing EM-GC modeled (*red*) and CRU observed (*black*) ΔT, as well as three of the terms that drive ΔT^{HUMAN} (Eq. 2.4) computed for the AerRF$_{2011}$ = −0.4 W m^{-2}, the IPCC (2013) upper limit of the likely range for ΔRF due to anthropogenic, tropospheric aerosols. The projection of ΔT to 2060 uses the indicated value of λ. The gold circles at 2060 are placed at the Paris target (1.5 °C) and upper limit (2.0 °C); (**b**) same as (**a**), except calculations conducted for AerRF$_{2011}$ = −0.9 W m^{-2}, the IPCC (2013) best estimate of ΔRF due to aerosols. Here, the contribution to ΔT from ENSO is also shown, so that the connection of anomalous warm conditions in 2015 to projected ΔT can be better visualized. The contribution of ENSO to ΔT is only shown once, since it is similar for all three simulations; (**c**) same as (**a**), except for AerRF$_{2011}$ = −1.5 W m^{-2}, the IPCC (2013) lower limit of the likely range for ΔRF due to anthropogenic, tropospheric aerosols. All calculations used the mean value of OHC computed from the six datasets shown in Fig. 2.8

(2) tropospheric aerosols have offset a considerable portion of the GHG warming over the prior decades because cooling (sulfate) has dominated heating (black carbon) and the **climate feedback** needed to fit observed ΔT_i is **large** (Fig. 2.9c).

If whatever value of climate feedback (model parameter λ) needed to fit the past climate record is assumed to be unchanged into the future, then projections of global warming under scenario 2 (Fig. 2.9c) far exceed those of scenario 1 (Fig. 2.9a). The fundamental reason for this dichotomy is that RF of climate due to all types of tropospheric aerosols will be much lower in the future than it has been in the past, due to public health legislation designed to improve air quality (Fig. 1.10). Future warming thus depends on ΔRF due to GHGs (same for both scenarios) and climate feedback (larger for scenario 2). When two different models can produce similarly good fits to a data record under contrasting assumptions, such as scenarios 1 and 2 above, physicists term the problem as being *degenerate*. Simply put, the degeneracy of Earth's climate introduces a fundamental uncertainty to projections of global warming.

The degeneracy of our present understanding of Earth's climate has important implications for policy. Figure 2.9 also contains markers, placed at year 2060, of the goal (1.5 °C warming) and upper limit (2.0 °C) of the Paris Climate Agreement. Again, all of the projections in Fig. 2.9 are based on RCP 4.5; the three simulations represent the present "likely" range of uncertainty in ΔRF of climate associated with the RCP 4.5 aerosol precursor specification. The projection of ΔT in Fig. 2.9a lies below the Paris goal for the entire time period; the projection of ΔT in Fig. 2.9b hits the Paris goal right at 2060, whereas the projection of ΔT in Fig. 2.9c falls between the Paris goal and upper limit in 2060. Later in this chapter we show projections out to year 2100, which is especially important since simulated temperatures are all rising at the end of the time period used for Fig. 2.9.

The calculations shown in Fig. 2.9 suggest that if the present uncertainty in ΔRF due to tropospheric aerosols could be reduced, then global warming could be projected more accurately. There is considerable effort in the climate community to reduce the uncertainty in this term. It is beyond the scope of this book to review the widespread efforts in this area; such reviews are the domain of large, community wide efforts such as the decadal surveys of measurement needs conducted by the US National Academy of Sciences (NAS).[20] Bond et al. (2013) published a detailed evaluation of the radiative effect due to black carbon (BC) aerosols and concluded the most likely value was 0.71 W m^{-2} warming, from 1750 to 2005, which far exceeds the IPCC (2007) estimate of 0.2 W m^{-2} warming over this same period of time. The IPCC (2013) best estimate of ΔRF for BC aerosols is 0.4 W m^{-2} warming, from 1750 to 2011. If the Bond et al. (2013) estimate is correct, then all else being equal, the absolute value of the best estimate for AerRF$_{2011}$ would drop, relative to the -0.9 W m^{-2} value given by IPCC (2013). Given the cantilevering between climate feedback and AerRF$_{2011}$ (Fig. 2.9) and the sensitivity of future ΔT to climate feedback, this modification would induce a corresponding decline in the associated

---

[20] At time of writing, the 2017 NAS Decadal Survey is underway and progress can be viewed at: http://sites.nationalacademies.org/DEPS/ESAS2017/index.htm

projection of ΔT. Much more work is needed to better quantify ΔRF due to aerosols, because of the complexity of aerosol types that affect the direct RF term (Kahn 2012) as well as difficulties in assessing the effect of aerosols on clouds (Morgan et al. 2006; Storelvmo et al. 2009).

2.2.1.4   Equilibrium Climate Sensitivity

The degeneracy of the climate record also limits our ability to precisely define equilibrium climate sensitivity (ECS), the warming that occurs after climate has equilibrated with $2 \times$ pre-industrial CO_2 (Kiehl 2007; Schwartz 2012; Otto et al. 2013). The values of ECS associated with the three simulations shown in Fig. 2.9 are 1.4, 1.7, and 2.4 °C, for $AerRF_{2011}$ values of -0.4 W m^{-2}, -0.9 W m^{-2}, and -1.5 W m^{-2}, respectively. We conclude from Fig. 2.9 that if ocean heat export occurs in a manner similar to that described by the OHC determined by averaging six data records, then ECS lies between 1.4 and 2.4 °C.

Alas, if only the climate system were this simple. As shown in Fig. 2.8, the OHC record is also quite uncertain. Figure 2.10 shows three additional simulations of Earth's climate, similar except for choice of OHC. All three simulations shown in Fig. 2.10 use the IPCC (2013) best estimate of -0.9 W m^{-2} for $AerRF_{2011}$. Figure 2.10a utilizes the OHC record of Ishii and Kimoto (2009), which yields the smallest value of κ among all available datasets, 0.43 W m^{-2} °C^{-1}. Figure 2.10c makes use of the OHC record of Gouretski and Reseghetti (2010) that yields the largest value of κ, 1.52 W m^{-2} °C^{-1}. The OHC record of Levitus et al. (2012), which lies closest to the average of the six OHC determinations (Fig. 2.8), results in an intermediate value of κ equal to 0.68 W m^{-2} °C^{-1} (Fig. 2.10b). The second rung of each ladder plot of Fig. 2.10 shows the contributions to ΔT^{HUMAN} from GHGs, tropospheric aerosols, and OHE.[21] The value of ECS ranges from 1.6 °C to 2.5 °C, depending on which dataset for OHC is used. These simulations reveal a ***second degeneracy*** of the climate record, which further impacts our ability to define ECS. If the export of heat from the atmosphere to the oceans is truly as large as suggested by the OHC record of Gouretski and Reseghetti (2010), then Earth's climate exhibits considerably larger sensitivity to the doubling of atmospheric CO_2 than if the OHC record of Ishii and Kimoto (2009) is correct.

Despite these complexities, an important pattern emerges upon comparison of ECS inferred from observations to ECS from GCMs. Figure 2.11 shows ECS from GCMs that had been used in IPCC (2007), the more recent IPCC (2013) GCMs, and a subset of the IPCC (2013) GCMs that participated in an evaluation process known as the Atmospheric Chemistry and Climate Model Intercomparison Project (ACCMIP). The ACCMIP GCMs tend to have more sophisticated treatment of tropospheric aerosols than the rest of the CMIP5 GCMs (Shindell et al. 2013). Figure 2.11 also shows three recent, independent estimates of ECS from the actual climate record: two based on analyses conceptually similar to our EM-GC approach, albeit quite different in design and implementation (Schwartz 2012; Masters 2014) and a

---

[21] The LUC term, which is always close to zero, is not shown in Fig. 2.10 for clarity.

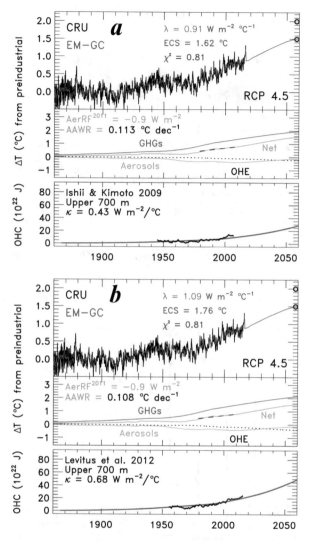

Fig. 2.10 (a) Observed and EM-GC simulated global warming, 1860–2015 as well as global warming projected to 2060. Top rung of a typical ladder plot, comparing EM-GC modeled (*red*) and CRU observed (*black*) ΔT, as well as three of the terms that drive ΔT^{HUMAN} (Eq. 2.4) computed for the AerRF$_{2011}$ = −0.9 W m^{-2}, the IPCC (2013) best estimate for ΔRF due to aerosols, and comparison of modeled and measured OHC, for a simulation that derives a value for κ that provides best fit to the OHC dataset of Ishii and Kimoto (2009). (b) Same as (a), expect for a simulation that derives a value for κ that provides best fit to the OHC dataset of Levitus et al. (2012). (c) Same as (a), expect for a simulation that derives a value for κ that provides best fit to the OHC dataset of Gouretski and Reseghetti (2010). Note how the values of Equilibrium Climate Sensitivity given in (a)–(c) respond to changes in OHC, whereas the transient climate response (*red curve*, upper rung of each ladder plot) are nearly identical. Also, smaller values of Attributable Anthropogenic Warming Rate (AAWR) are found as OHC rises, due to interplay of the OHE and aerosol terms within ΔT^{HUMAN}

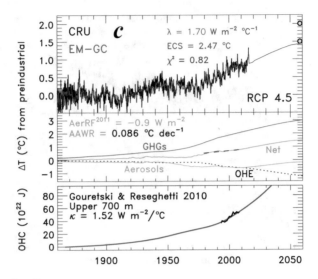

Fig.2.10 (continued)

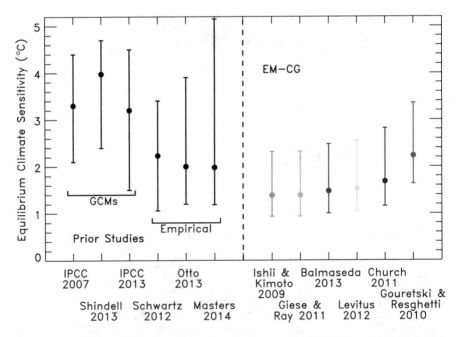

Fig. 2.11 Equilibrium Climate Sensitivity from the literature and EM-GC simulations. Estimates of ECS from six previously published studies (*left most points*, *black*) and from six runs of our Empirical Model of Global Climate (*right most points*, *colors*). For the six points to the left, words below the axis are the citation for the ECS value. For the six colored points to the right, the words below the axis denote the origin of the OHC record used in the particular EM-GC simulation. See Methods for further information

third that examined Earth's energy budget in detail over various decadal periods (Otto et al. 2013). The right hand side of Fig. 2.11 shows ECS found using our EM-GC framework, for the six estimates of OHC that appear in Fig. 2.8.

Figure 2.11 shows that published values of ECS from GCMs (average of the three best estimates is 3.5 °C) are considerably larger than estimates of ECS from the actual climate record. This pattern holds upon comparison of GCM-based ECS to values found using empirically-based estimates of ECS found by other research groups (mean value 2.1 °C) and using our EM-GC framework (mean value 1.6 °C).

These three estimates of ECS are important for policy. The mean value of ECS from GCMs (3.5 °C), taken literally and ignoring changes in other GHGs, indicates CO_2 must be kept far short of the 2 × pre-industrial level to achieve the Paris upper limit of 2 °C warming. The mean of the three empirically based estimates of ECS from other groups (2.1 °C) suggests the Paris upper limit can perhaps be achieved if the rise of CO_2 can be arrested before reaching the 2 × pre-industrial level, whereas the mean value ECS from our EM-GC framework (1.6 °C) suggests that if society manages to keep CO_2 from reaching 2 × pre-industrial level, the Paris goal might be achieved. Of course, these statements are all contingent on minimal future growth of other GHGs. Also, we stress that all of the estimates of ECS, even those from our EM-GC framework, are associated with considerable uncertainty. Nonetheless, the various ECS estimates in Fig. 2.11 suggest climate feedback within GCMs is larger than in the actual climate system,[22] which would explain the tendency for so many CMIP5 GCM projections of ΔT to lie above the green trapezoid in Fig. 2.3.

The tendency of CMIP5 GCMs to warm too quickly, with respect to the actual human influence on ΔT, is probed further in Sect. 2.3. This shortcoming of the CMIP5 GCMs is crucial to the thesis of this book: that the Paris Climate Agreement, as presently formulated, could actually limit the growth of GMST to less than 2 °C above pre-industrial.

2.3   Attributable Anthropogenic Warming Rate

The most important metric for a climate model is how well the prior rise in global mean surface temperature can be simulated. The green trapezoid used in various figures throughout this chapter is based on the recognition, by Chap. 11 of IPCC (2013), that CMIP5 GCMs have warmed too aggressively compared to observations over the prior several decades. In this section, the Empirical Model of Global Climate is used to quantify the amount of global warming that can be attributed to humans, over the time period 1979–2010.[23] These years are chosen because the rise in ΔT is nearly linear over this interval and this period has been the basis of similar examination by several other studies (Foster and Rahmstorf 2011; Zhou and Tung 2013). Our analysis of ΔT is compared to simulations of this quantity provided by CMIP5 GCMs, and to other analyses of ΔT over this period of time.

---

[22] Most estimates of ECS, such as Eq. 2.6, show ECS to be solely a function of climate feedback.

[23] Specifically all analyses in this section span the start of 1979 to the end of 2010.

First, some terminology must be defined. Chap. 10 of IPCC (2013) examined the amount of warming over specific time periods that can be attributed to humans, which we term Attributable Anthropogenic Warming (AAW). Figure 10.3 of IPCC (2013) shows plots of the latitudinal distribution of AAW, for time periods of 32, 50, 60, and 110 years. We prefer to divide AAW (units of °C) by the length of the time period in question, to arrive at a term called Attributable Anthropogenic Warming Rate (AAWR) (units of °C/decade). Consideration of AAWR, rather than AAW, provides a means to compare observed and modeled ΔT for studies that happen to examine time intervals with various lengths.

Next, the method for quantifying AAWR is described. Equation 2.4 provides a mathematical definition for ΔT^{HUMAN}_i in the EM-GC framework. This equation represents the contribution to the changes in GMST due to human release of GHGs, industrial aerosols, and land use change. Central to our estimate of AAWR is quantitative representation of the climate feedback needed to match observed ΔT (parameter γ in Eq. 2.4) and transfer of heat from the atmosphere to the ocean (term Q_{OCEAN}). The slope of ΔT^{HUMAN}_i found using Eq. 2.4, with respect to time, is used to define AAWR. Below, slopes are found by fitting values of ΔT^{HUMAN}_i for time periods that span the start of 1979 to the end of 2010, for various runs of the EM-GC that cover the entire 1860–2015 period of time.

Numerical values of AAWR, from 1979 to 2010, are recorded in Figs. 2.4, 2.5, 2.9, and 2.10. The uncertainty associated with each value of AAWR given in Figs. 2.4 and 2.5 is the standard error of the slope, found using linear regression.[24] The values of AAWR on these figures span a range of 0.086 °C/decade (Fig. 2.10c) to 0.122 °C/decade (Fig. 2.9c). Differences in AAWR reflect changes in the slope of ΔT^{HUMAN}_i over this 32-year interval, driven by various assumptions for ΔRF due to tropospheric aerosols as well as ocean heat export.

Figure 2.12 illustrates the dependence of AAWR on the specification of radiative forcing due to tropospheric aerosols. Panel b shows estimates of AAWR as a function of AerRF$_{2011}$, for simulations that all utilize the average value of ocean heat content from the six datasets shown in Fig. 2.8. The uncertainty of each data point represents the range of AAWR found for various assumptions regarding the shape of ΔRF of aerosols (i.e., the three curves for a specific value of AerRF$_{2011}$ shown in Fig. 2.7, all of which are tied to aerosol precursor emission files from RCP 4.5). Figure 2.12a shows the mean value of $1/\chi^2$ associated with the three simulations conducted for a specific value of AerRF$_{2011}$. The higher the value of $1/\chi^2$, the better the climate record is simulated. The best estimate for AAWR of 0.107 °C/decade is based on a weighted average of the five circles in Fig. 2.12b, where $1/\chi^2$ is used as the weight for each data point. The largest and smallest values of the five error bars in Fig. 2.12b are used to determine the upper and lower limits of AAWR, respectively. We conclude that if OHC has risen in a manner described by the average of the six datasets shown in Fig. 2.8, then the best estimate of AAWR over 1979–2010 is 0.107 °C/decade, with 0.080–0.143 °C/decade bounding the likely range.

The specific data record chosen for OHC has a modest effect on AAWR. This sensitivity is apparent from numerical values for AAWR recorded in Fig. 2.10a–c.

---

[24] Uncertainties for AAWR are omitted from Figs. 2.9 and 2.10, for clarity, but are of the same magnitude as the uncertainties given in Figs. 2.4 and 2.5.

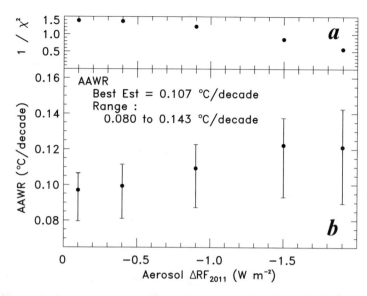

Fig. 2.12 Sensitivity of Attributable Anthropogenic Warming Rate to ΔRF of aerosols. (**a**) $1/\chi^2$ from the EM-GC simulations in the lower panel; the larger the value, the better the fit; (**b**) values of AAWR for 1979–2010, computed as the slope of ΔT^{HUMAN}, for EM-GC simulations that use the 15 time series of aerosol ΔRF shown in Fig. 2.7a. AAWR is displayed as a function of aerosol ΔRF in year 2011 (AerRF$_{2011}$). All calculations used the mean value of OHC computed from the six datasets shown in Fig. 2.8. The best estimate for AAWR, found using five estimates weighted by $1/\chi^2$, as well as the lower and upper estimates for AAWR, are indicated. See Methods for further information

This dependence of AAWR on OHC is illustrated by the colored symbols in Fig. 2.13, which show the best estimate (symbols) and range of AAWR (error bars) that is found for each of the six OHC records. The three groupings of data points show AAWR found using ΔT from CRU (Jones et al. 2012), GISS (Hansen et al. 2010), and NCEI (Karl et al. 2015). Nearly identical values of AAWR are found, regardless of which data center record is used to define ΔT. The mean value of the 18 empirical determinations of AAWR in Fig. 2.13 is 0.109 °C/decade, with a low and high of 0.028 and 0.170 °C/decade, respectively. The notation 0.109 (0.028, 0.170) °C/decade is used to denote the mean and range of this determination of AAWR.

Figure 2.13 also contains a graphical representation of AAWR extracted from the 41 GCMs that submitted results for RCP 4.5 to the CMIP5 archive (see Methods for details on how AAWR from GCMs is found). The GCM values of AAWR are displayed using a box and whisker symbol. The middle line represents the median value of AAWR from the GCMs; the box is bounded by the 25th and 75th percentiles, whereas the whisker (vertical line) connects the maximum and minimum values. The median value of AAWR from the CMIP5 GCMs is 0.218 °C/decade, about twice our best estimate of the actual rate of warming caused by human activities. The 25th percentile lies at 0.183 °C/decade, which exceeds the empirically determined upper limit for AAWR of 0.170 °C/decade over the time period 1979–2010. In other words, the CMIP5 GCMs on average simulate an anthropogenically induced rate of warming that is twice as fast as the actual climate system has warmed

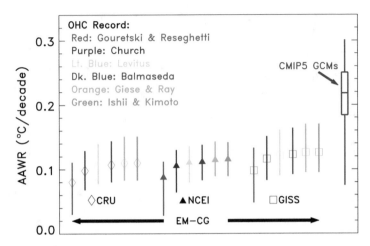

Fig. 2.13 Attributable Anthropogenic Warming Rate from the EM-GC and CMIP5 GCMs. Diamonds, triangles, and squares show the best estimate of AAWR, 1979–2010, found using ΔT from the CRU (Jones et al. 2012), GISS (Hansen et al. 2010), and NCEI (Karl et al. 2015) data centers, for various data records of OHC denoted by color. Error bars on these points represent the upper and lower limits of AAWR computed based on consideration of 15 possible time series for ΔRF of aerosols shown in Fig. 2.7a. Values of AAWR over 1979–2010 from the 41 GCMs that submitted RCP 4.5 simulations to the CMIP5 archive are shown by the box and whisker (BW) symbol. The middle line of the BW symbol shows the median value of AAWR from the 41 GCMs; the boxes denote the 25th and 75th percentile of the distribution, and the whiskers show maximum and minimum values of AAWR. See Methods for further information

and three quarters of the CMIP5 GCMs exhibit warming that exceeds the highest plausible value for AAWR that we infer from the climate record. This is rather disconcerting, given the prominence of the CMIP5 GCMs in the discussion of climate policy (e.g., Rogelj et al. 2016 and references therein).

The most likely reason for the shortcoming of CMIP5 GCMs illustrated in Fig. 2.13 is that climate feedback within these models is too large. Although tabulations of λ from CMIP5 GCMs exist (i.e., Table 9.5 of IPCC 2013), comparison to values of λ found using the EM-GC framework is complicated by the sensitivity of λ to the ΔRF of climate due to aerosols as well as ocean heat export. Most studies of GCM output (Shindell et al. 2013; Andrews et al. 2012; Vial et al. 2013) do not examine all three of these parameters. For meaningful comparison of GCMs to climate feedback from our simulations, it would be particularly helpful if future GCM tabulations of λ provided ΔRF due to aerosols and the ocean heat uptake efficiency coefficient (Raper et al. 2002) that best describes the rise ocean heat content within each GCM simulation. While the discussion of Fig. 9.17 of IPCC (2013) emphasizes good agreement between the observed rise in ocean heat content (OHC) and the CMIP5 multi-model mean rise in OHC since the early 1960s, there is an enormous range in the actual increase of OHC among the 27 CMIP5 GCMs used in their analysis.

Cloud feedback tends to be positive in nearly all GCMs; i.e., simulated changes in the properties and distribution of clouds tends to amplify ΔRF of climate due to rising

GHGs (Vial et al. 2013; Zelinka et al. 2013; Zhou et al. 2015).[25] Furthermore, GCMs that represent clouds in such a way that they act as a strong positive feedback tend to have larger values of ECS (Vial et al. 2013). It is quite challenging to define cloud feedback from observations because the effect of clouds on ΔRF of climate depends on cloud height, cloud thickness, and radiative effects in two distinct spectral regions.[26] To truly discern cloud feedback, the effect of anthropogenic tropospheric aerosols on clouds should be quantified and removed (Peng et al. 2016). The ephemeral nature of clouds requires either a long observing time to discern a signal from an inherently noisy process or the use of seasonal changes to deduce a relation between forcing and response (Dessler 2010). Nonetheless, evidence has emerged that cloud feedback in the actual atmosphere is indeed positive (Weaver et al. 2015; Zhou et al. 2015; Norris et al. 2016). However, the uncertainty in the empirical determination of cloud feedback is quite large (Dessler 2010; Zhou et al. 2015). Furthermore, the vast majority of satellite-based studies of cloud feedback that compare to GCM output make no attempt to quantify the effect of aerosols on clouds, which is problematic given the change in the release of aerosol precursors that has occurred in the past three decades (Smith and Bond 2014) combined with varied representation of the effect of aerosols on clouds within GCMs (Schmidt et al. 2014). There are major efforts underway to evaluate and improve the representation of clouds within GCMs (Webb et al. 2016). Based on the considerable existing uncertainty in the empirical determination of cloud feedback and the wide range of GCM representations of this process, cloud feedback within GCMs is the leading candidate for explaining why most of the GCM-based values of AAWR exceed the empirical determination of AAWR.

Next, our determination of AAWR is compared to estimates published by other groups. All studies considered here examined the time period 1979–2010. Our best estimate (and range) for AAWR found using the CRU ΔT dataset is 0.107 (0.080, 0.143) °C/decade. Foster and Rahmstorf (2011) (hereafter, FR2011) reported a value for AAWR of 0.170 °C/decade based on analysis of an earlier version of the CRU ΔT record.[27] They used multiple linear regression to remove the influence of ENSO, volcanoes, and total solar irradiance on observed ΔT and then examined the difference between observed ΔT and the contribution from these three exogenous factors, termed the residual, to quantify ΔT. The FR2011 estimate of AAWR exceeds the upper limit of our analysis shown in Fig. 2.12 and lies closer to median GCM-based value of 0.218 °C/decade found upon our analysis of the CMIP5 archive.

The difference between our best estimate for AAWR (0.107 °C/decade) and the value reported by FR2011 (0.170 °C/decade), both for ΔT from CRU, is due to the two approaches used to quantify the human influence on global warming. We have applied

---

[25] Figure 7.10 of IPCC (2013) provides a concise summary of the representation of cloud feedback within GCMs.

[26] Proper determination of ΔRF due to clouds requires analysis of the impact of clouds on reflectivity and absorption of solar radiation, commonly called the cloud short wavelength (SW) effect in the literature, as well as the impact of clouds on the trapping of infrared radiation (or heat) emitted by Earth's surface, commonly called the long wavelength (LW) effect.

[27] FR2011 also reported slightly higher values of AAWR, 0.171 and 0.175 °C/decade, upon use of ΔT from GISS and NCEI, respectively.

the approach of FR2011 to the derivation of AAWR using both the older version of the CRU ΔT used in their study and the more recent version used in our analysis, and arrive at 0.166 °C/decade for the older version and 0.183 for the latest version.

The difficulty in the approach used by FR2011 is that their value of AAWR is based upon analysis of a residual found upon removal of all of the natural processes thought to influence ΔT. If an unaccounted for natural processes happens to influence ΔT over the period of time upon consideration, such as the Atlantic Meridional Overturning Circulation, then the value of AAWR found by examination of the residual will be biased by the magnitude of the variation in ΔT due to this process over the period of time under consideration.

Quantitative analysis of the CRU data record reveals the cause of the difference of these two apparently disparate estimates of AAWR for the 1979–2010 time period. The fifth rung of the Fig. 2.5 ladder plot indicates AMOC may have contributed 0.043 °C/decade to the rise of ΔT, over the time period 1979–2010. Upon use in our EM-GC framework of the same version of CRU ΔT that was analyzed by FR2011, we compute AAWR = 0.109 °C/decade and a slope of 0.058 °C/decade for the contribution of AMOC to ΔT over 1979–2010. Thus, natural variation of climate due to variations in the strength of the Atlantic Meridional Overturning Circulation accounts, nearly exactly, for the difference between the FR2011 estimate of AAWR (0.170 °C/decade) and our value (0.109 °C/decade).[28]

There is considerable debate about whether North Atlantic SST truly provides a proxy for variations in the strength of AMOC. An independent analysis conducted using different methodology (DelSole et al. 2011) supports our view that internal climate variability contributed significantly to the relative warmth of latter part of the time series examined by FR2011. Analysis of a residual to quantify a process, rather than construction and application of a model that physically represents the process, violates fundamental principles of separation of signal from noise (Silver 2012). The estimates of AAWR shown in Figs. 2.4 and 2.5 yield *similar values*, 0.111 °C/decade versus 0.109 °C/decade, whether or not AMOC is considered, because our determination of AAWR is built upon a *physical model for the human influence on climate* (Eq. 2.4) and does not rely on analysis of a residual.

If there is one word that best summarizes the present state of climate science in the published literature, it might be confusion. Alas, the argument put forth in the prior paragraphs, that a value for AAWR from 1979 to 2010 of ~0.10 °C/decade is inferred from the climate record whether or not variations in the strength of AMOC are considered in the model framework, is in direct contradiction to Zhou and Tung (2013) (hereafter ZT2013). ZT2013 examined version 4 of the CRU ΔT data record, using a modified residual method,[29] and concluded AAWR is 0.169 °C/decade if temporal variation of AMOC is ignored, but drops to 0.07 °C/decade if variations in

---

[28] That is, 0.109 + 0.058 °C/decade is nearly equal to 0.170 °C/decade.

[29] The method used by ZT13 is similar to that of FR2011, except ZT13 include a model for ΔT^{HUMAN} in their calculation of regression coefficients that are used to remove the influence of ENSO, volcanic, and solar variations from ΔT (their case 1) or remove the influence of ENSO, volcanic, solar variations, and AMOC from ΔT (their case 2). For both cases, their model of ΔT^{HUMAN} is a linear function from 1860 to 2010.

the strength of AMOC are considered. The ZT13 estimate of AAWR without consideration of AMOC is in close agreement with the value published by FR2011, and disagrees with our value for the reasons described above.

The importance of the ZT13 study is that if their value of AAWR found upon consideration of AMOC (0.07 °C/decade) is correct, one would conclude that the CMIP5 GCMs warm a factor of three more quickly than the actual climate system has responded to human influence. We are also able to reproduce the results of ZT13, but we argue their estimate of AAWR is biased low because they used a single linear function to describe ΔT^{HUMAN} over the entire 1860–2010 time period. As illustrated on the second rung of the Figs. 2.4 and 2.5 ladder plots, ΔT^{HUMAN} varied in a non-linear manner from 1860 to present. The time variation of ΔT^{HUMAN} bears a striking resemblance to the rise in population over this period of time. For the determination of AAWR, not only should a model for ΔT^{HUMAN} be used, but this model must correspond to the actual shape of the time variation of radiative forcing of climate caused by humans.

2.4 Global Warming Hiatus

The evolution of ΔT over the time period 1998–2012 has received enormous attention in the popular press, blogs, and scientific literature because some estimates of ΔT over this period of time indicate little change (Trenberth and Fasullo 2013). Various suggestions had been put forth to explain this apparent leveling off of ΔT, including climate influence of minor volcanoes (Schmidt et al. 2014; Santer et al. 2014; Solomon et al. 2011), changes in ocean heat uptake (Balmaseda et al. 2013; Meehl et al. 2011), and strengthening of trade winds in the Pacific (England et al. 2014). The major ENSO event of 1998, which led to a brief, rapid rise in ΔT due to suppression of the upwelling of cold water in the eastern Pacific, must be factored into any analysis of the hiatus.[30]

Karl et al. (2015) have questioned the existence of a hiatus. They showed an update to the NCEI record of GMST, used to define ΔT, which exhibits a steady rise from 1998 to 2012, despite the ENSO event in 1998. The main improvement was extension to present time of a method to account for biases in SST, introduced by varying techniques to record water temperature from ship-borne instruments.

Figure 2.14 compares measured ΔT over 1998–2012 to simulations of ΔT from the EM-GC. The EM-GC simulations were conducted for the entire 1860–2015 time period: the figure zooms in on the time period of interest. Figure 2.14a–c shows results using the latest version of ΔT from CRU, GISS, and NCEI (footnotes 1 to 3 provide URLs, data versions, etc.). Each panel also includes the slopes of a linear fit to the data (black) and to modeled ΔT (red), over 1998–2012.

For the first time in our extensive analysis, the choice of a data center for ΔT actually matters. The observed time series of ΔT from CRU in Fig. 2.14 exhibits a

---

[30] The effect of ENSO on ΔT in 1998 is readily apparent on the fourth rung of Figs. 2.4 and 2.5 ladder plots.

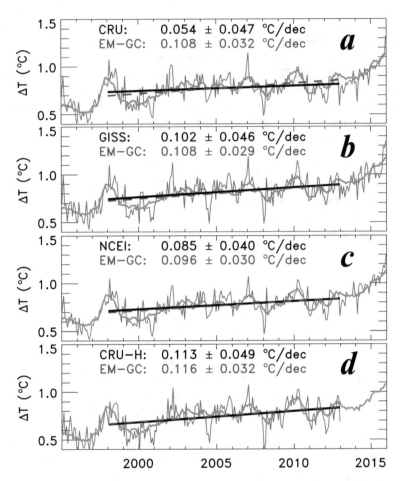

Fig. 2.14 Observed and EM-GC simulated ΔT, 1995–2016. Top rung of a typical ladder plot, comparing EM-GC modeled (*red*) and observed (*grey*) ΔT. Also shown are linear fits to the modeled (*red dashed*) and measured (*black*) time series of ΔT, considering monthly values from the start of 1998 to the end of 2012. The slope and standard error of each slope are also recorded. (**a**) ΔT from CRU was used (Jones et al. 2012); (**b**) ΔT from GISS (Hansen et al. 2010); (**c**) ΔT from NCEI (Karl et al. 2015); (**d**) ΔT from the CRU Hybrid adjustment of Cowtan and Way (2014). The linear fits to modeled ΔT for NCEI and CRU-H lie right on top of the respective fits to measured ΔT

slope of 0.054 ± 0.05 °C/decade over this 15-year period, about a factor of two less than the modeled slope of 0.108 ± 0.03 °C/decade. These two slopes do agree within their respective uncertainties and, as is visually apparent, the ~155-year long simulation does capture the essence of the observed variations reported by CRU over the time period of the so-called hiatus. Nonetheless, the slopes disagree by a factor of 2, lending credence to the idea that some change in the climate system not picked up by the EM-GC approach could be responsible for a gap between the modeled and measured ΔT between 1998 and 2012.

Analysis of the GISS and NCEI data sets leads to a different conclusion. As shown in Fig. 2.14b, c, the observed and modeled slope of ΔT, for 1998–2012, agree extremely well. The GISS record of GMST is based on the same SST record used by NCEI. Earlier versions of the NCEI record (not shown), released prior to the update in SST described by Karl et al. (2015), did support the notion that some unknown factor was suppressing the rise in ΔT from 1998 to 2012.

Cowtan and Way (2014) (hereafter, CW2014) suggest the existence of a recent, cool bias in the CRU estimate of ΔT, due to closure of observing stations in the Arctic and Africa that they contend has not been handled properly in the official CRU data release. CW2014 published two alternate versions of the CRU data set, termed "kriging" and "hybrid", to account for the impact of these station closures on ΔT. Figure 2.14d shows that, upon use of the CRU-Hybrid data set of CW2014, the observed and modeled slope of ΔT are in excellent agreement. Similarly good agreement between measured and modeled ΔT is obtained for CRU-Kriging (not shown). It remains to be seen whether CW2014 will impact future versions of ΔT from CRU. In the interim, the CW2014 analysis supports the finding, from the GISS and NCEI data sets, that there was no hiatus in the gradual, long-term rise of ΔT.

The EM-GC allows us to extract AAWR for any period of time. For the simulations shown in four panels of Fig. 2.14, the values of AAWR for 1998–2012 are 0.1075 ± 0.0041, 0.1186 ± 0.004, 0.1089 ± 0.0046, and 0.1039 ± 0.004, respectively, all in units of °C/decade. The primary factors responsible for the slightly smaller rise in ΔT (black numbers, Fig. 2.14) compared to AAWR over 1998–2012 is the tendency of the climate system to be in a more La Niña like state during the latter half of this period of time[31] (Kosaka and Xie 2013) and a relatively small value of total solar irradiance during the most recent solar max cycle (Coddington et al. 2016). Our simulations, which include Kasatochi, Sarychev and Nabro, suggest these recent minor volcanic eruptions played only a miniscule role (~0.0018 °C/decade cooling) over this period. We conclude human activity exerted about 0.11 °C/decade warming over 1998–2012, and observations show a rise of ΔT that is slightly smaller in magnitude, due to natural factors that are well characterized by the Empirical Model of Global Climate.

2.5 Future Temperature Projections

Accurate projections of the expected future rise of GMST are central for the successful implementation of the Paris Climate Agreement. As shown in Sect. 2.2.1.3, the degeneracy of the climate system coupled with uncertainty in ΔRF due to tropospheric aerosols leads to considerable spread in projections of ΔT (the anomaly of

---

[31] This is not particularly surprising given the strong ENSO of 1998. Hindsight is 20:20, but it is nonetheless remarkable how much attention has been devoted to discussion of ΔT over the 1998–2012 time period, including within IPCC (2013), given the unusual climatic conditions known to have occurred at the start of this time period. Apparently the global warming deniers took the lead in promulgating the notion that more than a decade had passed without a discernable rise in ΔT, and the scientific community took that bait and devoted enormous resources to examination of GMST over this particular 15-year interval.

GMST relative to pre-industrial background). Complicating matters further, CMIP5 GCMs on average overestimate the observed rate of increase of ΔT during the past three decades by about a factor of two (Sect. 2.3). Recognition of the tendency of CMIP5 GCMs to overestimate observed ΔT led Chap. 11 of IPCC (2013) to issue a revised forecast for the rise in GMST over the next two decades, which is featured prominently below. Here, these issues are briefly reviewed in the context of the projections of ΔT relevant for evaluation of the Paris Climate Agreement. Finally, a route forward is described, based on forecasts of ΔT from the Empirical Model of Global Climate (EM-GC) (Canty et al. 2013).

Figure 2.15 provides dramatic illustration of the impact on global warming forecasts of the degeneracy of Earth's climate system. These so-called ellipse plots show calculations of ΔT in year 2060 (ΔT_{2060}) (various colors) computed using the EM-GC, as a function of model parameters λ (climate feedback) and $AerRF_{2011}$ (ΔRF due to tropospheric aerosols in year 2011). Estimates of ΔT_{2060} are shown only if a value of $\chi^2 \leq 2$ can be achieved for a particular combination of λ and $AerRF_{2011}$. In other words, the ellipse-like shape of ΔT_{2060} defines the range of these model parameters for which an acceptable fit to the climate record can be achieved. The EM-GC simulations in Fig. 2.15a utilize forecasts of GHGs and aerosols from RCP 4.5 (Thomson et al. 2011), whereas Fig. 2.15b is based on RCP 8.5 (Riahi et al. 2011). As noted above, projections of ΔT consider only human influences. We limit ΔRF due to aerosols to the possible range of IPCC (2013): i.e., $AerRF_{2011}$ must lie between -0.1 and -1.9 W m^{-2}. Even though values of $\chi^2 \leq 2$ can be achieved for values of λ and $AerRF_{2011}$ outside of this range, the corresponding portion of the ellipse is shaded grey and values of ΔT associated with this regime of parameter space are not considered. Projections of ΔT are insensitive to which OHC data record is chosen (Fig. 2.10), but the location of the ellipse on analogs to Fig. 2.15 varies, quite strongly in some cases, depending on which OHC data set is used. The $\chi^2 \leq 2$ ellipse-like feature upon use of OHC from Gouretski and Reseghetti (2010) is associated with larger values of λ than the ellipses that appear in Fig. 2.15; conversely, the ellipse-like feature found using OHC from Ishii and Kimoto (2009) is aligned with smaller values of λ. In both cases, the numerical values of ΔT_{2060} within the resulting ellipses are similar to those shown in Fig. 2.15.

Figure 2.16 is similar to Fig. 2.15, except projections of ΔT for year 2100 (ΔT_{2100}) are shown. The range of ΔT associated with the acceptable fits is recorded on all four panels of Fig. 2.15 and 2.16. For RCP 4.5, projected ΔT lies between 0.91 and 2.28 °C in 2060 and falls within 0.91 and 2.40 °C in 2100. This large range for projections of ΔT is quite important for policy, given the Paris goal and upper limit of restricting ΔT to 1.5 °C and 2.0 °C above the pre-industrial level, respectively. The large spread in ΔT is due to the degeneracy of our present understanding of climate. In other words, the climate record can be fit nearly equally well assuming either:

(1) Small aerosol cooling (values of $AerRF_{2011}$ close to -0.4 W m^{-2}) and weak climate feedback, which is associated with lower values of ΔT_{2060}.

(2) Large aerosol cooling (values of $AerRF_{2011}$ close to -1.5 W m^{-2}) and strong climate feedback, which is associated with higher values of ΔT_{2060}.

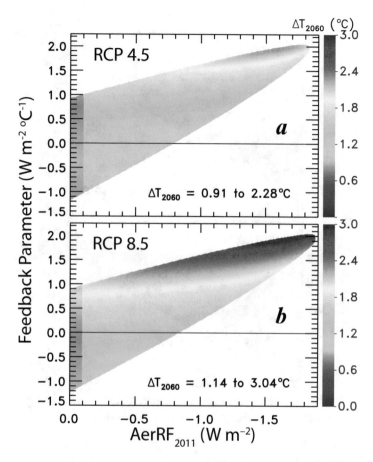

Fig. 2.15 Projected rise in GMST, year 2060, as a function of climate feedback and aerosol radiative forcing. Values of ΔT relative to the pre-industrial baseline found using the EM-GC framework, for all combinations of model parameters λ and $AerRF_{2011}$ that provide an acceptable fit to the climate record, defined here as yielding a value of $\chi^2 \leq 2$. Projections of ΔT are shown only for $AerRF_{2011}$ between the IPCC (2013) limits of -1.9 and -0.1 W m^{-2}. The color bar denotes ΔT_{2060} found by considering only the ΔT^{HUMAN} term in Eq. 2.2 for the future. All simulations used OHC from the average of six data records shown in Fig. 2.8 and the aerosol ΔRF time series are based on scaling parameters along the middle road of Fig. 2.21. (**a**) GHG and aerosol ΔRF based on RCP 4.5 (Thomson et al. 2011); (**b**) GHG and aerosol ΔRF based on RCP 8.5 (Riahi et al. 2011). The minimum and maximum values of ΔT_{2060} are recorded on each panel

Studies of tropospheric aerosol ΔRF are unable, at present time, to definitely rule out any of these possibilities.

One clear message that emerges from Figs. 2.15 and 2.16 is that to achieve the goal of the Paris Climate Agreement, emissions of GHGs must fall significantly below those used to drive RCP 8.5. The range of ΔT_{2100} shown in Fig. 2.16b is 1.6–4.7 °C. Climate catastrophe (rapid rise of sea level, large shifts in patterns of drought and flooding, loss of habitat, etc.) will almost certainly occur by end of this

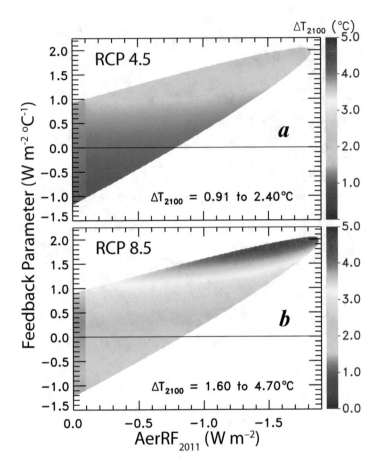

Fig. 2.16 Projected rise in GMST, year 2100, as a function of climate feedback and aerosol radiative forcing. Same as Fig. 2.15, except for EM-GC projections out to year 2100. The same color bar is used for both panels to accentuate the end of century difference between RCP 4.5 and RCP 8.5. The minimum and maximum values of ΔT_{2100} are recorded on each panel

century if the emissions of GHGs, particularly CO_2, follow those used to drive RCP 8.5.[32] The book *Six Degrees: Our Future on a Hotter Planet* (Lynas 2008) provides an accessible discourse of the consequences of global warming, organized into 1 °C increments of future ΔT.

In the rest of this chapter, policy relevant projections of ΔT are shown, both from the EM-GC framework and CMIP5 GCMs. Figures 2.17 shows the statistical distribution of ΔT_{2060} from our EM-GC calculations. The EM-GC based projections are weighted by $1/\chi^2$ (i.e., the better the fit to the climate record, the more heavily a particular projection is weighted). The height of each histogram represents the probability that a particular range of ΔT_{2060}, defined by the width of each line segment,

---

[32] As shown in Fig. 2.1, CO_2 and CH_4 reach alarmingly high levels at end of century in the RCP 8.5 scenario.

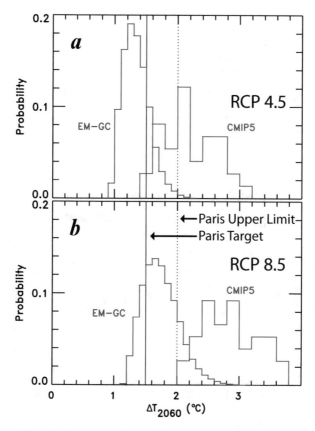

Fig. 2.17 Probability distribution functions of rise in GMST in year 2060. The line segments represent a series of histograms (*narrow, vertical rectangles*) for projections of ΔT in year 2060 relative to the pre-industrial baseline found using our EM-GC (*blue*) and CMIP5 GCMs (*red*). The height of each histogram represents the probability the rise of ΔT will fall within the range of ΔT that corresponds to the ends of each line segment (see main text). The Paris Climate Agreement target and upper limit of 1.5 and 2.0 °C warming are denoted. Projections of ΔT_{2060} found using the EM-GC consider only combinations of model parameters λ and $AerRF_{2011}$ that fall within the respective ellipse of Fig. 2.17 (i.e., projections consider only acceptable fits to the climate record) and the EM-GC values of ΔT_{2060} are weighted by $1/\chi^2$, so that simulations that provide a better fit to the climate record are given more credence. Finally, the EM-GC simulations used OHC from the average of six data records shown in Fig. 2.8 and the aerosol ΔRF time series based on scaling parameters along the middle road of Fig. 2.21. (**a**) EM-GC and CMIP5 GCM projections based on RCP 4.5 (Thomson et al. 2011); the GCM projections consider the 41 models represented in Fig. 2.3a ; (**b**) EM-GC and CMIP5 GCM projections based on RCP 8.5 (Riahi et al. 2011); the GCM projections consider the 38 models represented in Fig. 2.3b

will occur. In other words, the most probable value of ΔT in year 2060, for the EM-GC projection that uses RCP 4.5, is 1.2–1.3 °C above pre-industrial, and there is slightly less than 20 % probability ΔT will actually fall within this range. In contrast, the CMIP5 GCMs project ΔT in 2060 will most probably be 2.0–2.2 °C warmer than pre-industrial, with a ~12 % probability ΔT will actually fall within this range. A finer spacing for ΔT is used for the EM-GC projection, since we are

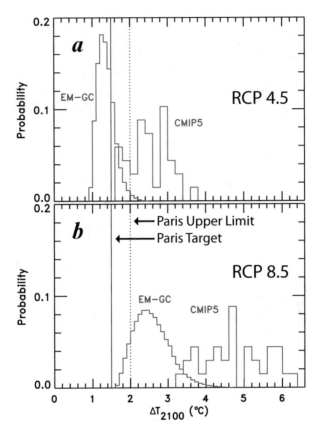

Fig. 2.18 Probability distribution functions of rise GMST, year 2100. Same as Fig. 2.17, except all of the projections are for year 2100

able to conduct many simulations in this model framework. Figure 2.18 is similar to Fig. 2.17, except the projection is for year 2100. The collection of histograms shown for any particular model (i.e., either CMIP5 GCMs or EM-GC) on a specific figure is termed the probability distribution function (PDF) for the projection of the rise in GMST (i.e., ΔT).

The PDFs shown in Figs. 2.17 and 2.18 reveal stark differences in projections of ΔT based on the EM-GC framework and the CMIP5 GCMs. In all cases, ΔT from the GCMs far exceed projections using our relatively simple approach that is tightly coupled to observed ΔT, OHC, and various natural factors that influence climate. These differences are quantified in Table 2.1, which summarizes the cumulative probability that a specific Paris goal can be achieved. The cumulative probabilities shown in Table 2.1 are based on summing the height of each histogram that lies to the left of a specific temperature, in Figs. 2.17 and 2.18.

Time series of ΔT found using the CMIP5 GCM and EM-GC approaches are illustrated in Figs. 2.19 and 2.20, which show projections based on RCP 4.5 and

Table 2.1 Cumulative probability the rise in ΔT remains below a specific value, 2060 and 2100

	2060		2100	
	1.5 °C	2.0 °C	1.5 °C	2.0 °C
CMIP5 GCMs RCP 4.5	0.027	0.270	0.0	0.206
CMIP5 GCMs RCP 8.5	0.0	0.026	0.0	0.0
EM-GC, RCP 4.5	0.787	0.995	0.751	0.989
EM-GC, RCP 8.5	0.215	0.816	0.0	0.098

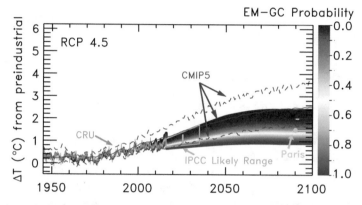

Fig. 2.19 Global warming projections, RCP 4.5. Simulations of the GMST anomaly relative to pre-industrial baseline (ΔT), found using the EM-GC (*red, white,* and *blue* colors) and from the CMIP5 GCMs (*grey lines*). Observed ΔT from CRU is also shown (*orange*). All simulations extend back to 1860; the figure shows ΔT from 1945 to 2100 so that the projections can be better visualized. The *green trapezoid* shows the indicative likely range of annual average ΔT for 2016 to 2035 (*roof and base of trapezoid are upper and lower limits*) and the green bar indicates the likely range of the mean value of ΔT over 2006 to 2035, both given in Chap. 11 of IPCC (2013). The Paris Climate Agreement target and upper limit of 1.5 and 2.0 °C warming are denoted at the end of the century. The three CMIP5 lines represent the minimum, maximum, and multi-model mean of ΔT from the 41 GCMs that submitted projections for RCP 4.5 to the CMIP5 archive. The EM-GC projections represent the probability that future value of ΔT will rise to the indicated level. As for Fig. 2.17, EM-GC projections consider only acceptable fits to the climate record, are based on the average of OHC from six data records, and have been weighted by $1/\chi^2$ prior to calculation of the probabilities. The white patch of the red, white, and blue projection is the most probable future value of ΔT found using this approach

RCP 8.5. The colors represent the probability of a particular future value of ΔT being achieved, for projections computed in the EM-GC framework weighted by $1/\chi^2$. Essentially, the red (warm), white (mid-point), and blue (cool) colors represent the visualization of a succession of histograms like those shown in Figs. 2.17 and 2.18. The GCM CMIP5 projections of ΔT (minimum, maximum, and multi-model mean) for RCP 4.5 and RCP 8.5 are shown by the three grey lines. These lines, identical to those shown in Fig. 2.3a (RCP 4.5) and Fig. 2.3b (RCP 8.5), are based on our analysis of GCM output preserved on the CMIP5 archive. The green trapezoid, which originates from Fig. 11.25b of IPCC (2013), makes a final and rather

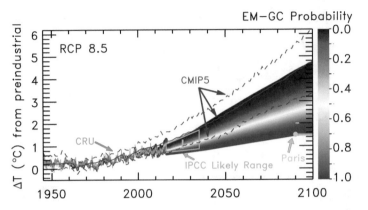

Fig. 2.20 Global warming projections, RCP 8.5. Same as Fig. 2.19, except for the 38 GCMs that submitted projections using RCP 8.5 to the CMIP5 archive. Note how the most probable evolution of ΔT found using the EM-GC framework passes through the middle of the IPCC (2013) trapezoid, and is matched only by the lowest projection warmings of the CMIP5 GCMs

important appearance on these figures. Also, the Paris target (1.5 °C) and upper limit (2 °C) are marked on the right vertical axis of both figures.

There are resounding policy implications inherent in Figs. 2.17, 2.18, 2.19, and 2.20. First, most importantly, and beyond debate of any reasonable quantitative analysis of climate, if GHG emissions follow anything close to RCP 8.5, there is no chance of achieving either the goal or upper limit of the Paris climate agreement (Fig. 2.20). Even though there is a small amount of overlap between the Paris targets and our EM-GC projections for year 2100 in Fig. 2.20, this is a false hope. In the highly unlikely event this realization were to actually happen, it would just be a matter of time before ΔT broke through the 2 °C barrier, with all of the attendant negative consequences (Lynas 2008). Plus, of course, 1.5–2.0 °C warming (i.e., the lead up to breaking the 2 °C barrier) could have rather severe consequences. This outcome is all but guaranteed if GHG abundances follow that of RCP 8.5.

The second policy implication is that projections of ΔT found using the EM-GC framework indicate that, if emissions of GHGs can be limited to those of RCP 4.5, then by end-century there is:

(a) a 75 % probability the Paris target of 1.5 °C warming above pre-industrial will be achieved
(b) a greater than 95 % probability the Paris upper limit of 2 °C warming will be achieved

As will be shown in Chap. 3, the cumulative effect of the commitments from nations to restrict future emissions of GHGs, upon which the Paris Climate Agreement is based, have the world on course to achieve GHG emissions that fall just below those of RCP 4.5, provided: (1) both conditional and unconditional commitments are followed; (2) reductions in GHG emissions needed to achieve the Paris agreement, which generally terminate in 2030, are continually improved out to at least 2060.

The policy implication articulated above differs considerably from the consensus in the climate modeling community that emission of GHGs must follow RCP 2.6 to achieve even the 2 °C upper limit of Paris (Rogelj et al. 2016). We caution those quick to dismiss the simplicity of our approach to consider the emerging view, discussed in Chap. 11 of IPCC (2013) and quantified in their Figs. 11.25 and TS.14, as well as our Figs. 2.3 and 2.13, that the CMIP5 GCMs warm much quicker than has been observed during the past three decades. In support of our approach, we emphasize that our projections of ΔT are bounded ***nearly exactly*** by the green trapezoid of IPCC (2013), which reflects the judgement of at least one group of experts as to how ΔT will evolve over the next two decades. Given our present understanding of Earth's climate system, we contend the Paris Climate Agreement is a beacon of hope because it places the world on a course of having a reasonable probability of avoiding climate catastrophe.

We conclude by cautioning against over-interpretation of the numbers in Table 2.1 or the projections in Figs. 2.19 and 2.20. Perhaps the largest source of uncertainty in the EM-GC estimates of ΔT is the assumption that whatever values of λ (climate feedback) and κ (ocean heat export coefficient) have occurred in the past will continue into the future. Should climate feedback rise, or ocean heat export fall, the future increase of ΔT will exceed that found using our approach. On the other hand, the past climate record can be fit exceedingly well for time invariant values of λ and κ. The great difficulty is that the specific values of these two parameters are not able to be ascertained from the climate record, due to large current uncertainties in ΔRF due to aerosols and the ocean heat content record. Community-wide efforts to reduce the uncertainties in ΔRF of aerosols and ocean heat storage are vital. We urge that judgement of the veracity of the results of our EM-GC projections be based on whether other research groups are able to reproduce these projections of ΔT, based on similar types of analyses. Given these caveats, our forecasts of global warming suggest that GHG emissions of RCP 4.5 constitute a reasonable guideline for attempting to achieve the both the Paris target (1.5 °C) and upper limit (2.0 °C) for global warming, relative to the pre-industrial era.

2.6 Methods

Many of the figures use data or archives of model output from publically available sources. Here, webpage addresses of these archives, citations, and details regarding how data and model output have been processed are provided. Only those figures with "see methods for further information" in the caption are addressed below. Electronic copies of the figures are available on-line at http://parisbeaconofhope.org.

Figure 2.1 shows mixing ratios of CO_2, CH_4, and N_2O from RCP 2.6, RCP 4.5, RCP 6.0, and RCP 8.5, which were obtained from files:

RCP*MIDYEAR_CONCENTRATIONS.DAT provided by the Potsdam Institute for Climate Research (PICR) at: http://www.pik-potsdam.de/~mmalte/rcps/data

The figures also contain observed global, annually averaged mixing ratios for each GHG. Observed CO_2 is from data provided by NOAA Earth Science Research Laboratory (ESRL) (Ballantyne et al. 2012) at: ftp://ftp.cmdl.noaa.gov/products/trends/co2/co2_annmean_gl.txt

The CO_2 record given at the above URL starts in 1980. This record has been extended back to 1959 using annual, global average CO_2 growth rates at: http://www.esrl.noaa.gov/gmd/ccgg/trends/global.html#global_growth

The CH_4 record for 1984 to present (Dlugokencky et al. 2009) is from: ftp://aftp.cmdl.noaa.gov/products/trends/ch4/ch4_annmean_gl.txt

For years prior to 1984, CH_4 is from a global average computed based on measurements at the Law Dome (Antarctica) and Summit (Greenland) ice cores (Etheridge et al. 1998): http://cdiac.ornl.gov/ftp/trends/atm_meth/EthCH498B.txt

The N_2O record for 1979 to present (Montzka et al. 2011) is from: ftp://ftp.cmdl.noaa.gov/hats/n2o/combined/HATS_global_N2O.txt

Figure 2.2 shows ΔRF of climate due to GHGs, for RCP 4.5 and RCP 8.5. The GHG abundances all originate from the files provided by PICR given for Fig. 2.1. The estimates of ΔRF for each GHG other than tropospheric O_3 were found using formulae in Table 8.SM.1 of IPCC (2013), which are identical to formulae given in Table 6.2 of IPCC (2001) except the value for pre-industrial CH_4 has risen from 0.700 to 0.722 ppm. These formulae use 1750 as the pre-industrial initial condition, as has been the case in all IPCC reports since 2001. Hence, ΔRF represents the increase in radiative forcing of climate since 1750. Throughout this book, we relate ΔRF computed in this manner to ΔT relative to a pre-industrial baseline of 1850–1900. This mismatch of baseline values for ΔRF and ΔT is a consequence of the IPCC precedent of initializing ΔRF in 1750 combined with 1850 marking the first thermometer based estimate of GMST provided by the Climate Research Unit of East Anglia, UK (Jones et al. 2012). The rise in RF of climate between 1750 and 1900 was small, so the mismatch of baselines has no significant influence on our analysis. The ΔRF due to tropospheric O_3 is based on the work of Meinshausen et al. (2011), obtained from the PICR files. The grouping of GHGs into various categories in Fig. 2.2 is the same as used for Fig. 1.4.

Figure 2.3 shows time series of ΔT, relative to the pre-industrial baseline, from CRU (Jones et al. 2012), GISS (Hansen et al. 2010), and NCEI (Karl et al. 2015) as well as GCMs that submitted model results to the CMIP5 archive (Taylor et al. 2012) for RCP 4.5 (Fig. 2.3a) and RCP 8.5 (Fig. 2.3b). The URLs of observed ΔT are given in footnotes 1, 2, and 3. The CMIP5 URL is given in footnote 5.

All of the observed ΔT time series are normalized to a baseline for 1850–1900 in the following manner. The raw CRU dataset is provided for a baseline of 1961–1990; the raw GISS dataset is provided for a baseline of 1951–1980, and the raw NCEI time series for ΔT is given relative to baseline of 1901–2000. The CRU dataset starts in 1850; the other two time series start in 1880. To transform each time series so that ΔT is relative to 1850–1900, the following steps are taken:

(a) for CRU, 0.3134 °C is added to each value of ΔT; 0.3134 °C is the difference between the mean of CRU ΔT during 1961–1990 relative to 1850–1900;

Table 2.2 Names of the 42 CMIP5 GCMs used in Fig. 2.3

1. ACCESS1.0	22. GFDL-ESM2M
2. ACCESS3.0	23. GISS-E2-H
3. BCC-CSM1.1	24. GISS-E2-H-CC
4. BCC-CSM1.1(m)	25. GISS-E2-R
5. BNU-CSM	26. GISS-E2-R-CC
6. CCSM4	27. HadCM3
7. CESM1(BGC)	28. HadGEM2-CC
8. CESM1(CAM5)	29. HadGEM2-ES
9. CMCC-CESM	30. INM-CM4
10. CMCC-CM	31. IPSL-CM5A-LR
11. CMCC-CMS	32. IPSL-CM5A-MR
12. CNRM-CM5	33. IPSL-CM5B-LR
13. CSIRO-Mk3.6.0	34. MIROC-ESM
14. CanCM4	35. MIROC-ESM-CHEM
15. CanESM2	36. MIROC4h
16. EC-EARTH	37. MIROC5
17. FGOALS-g2	38. MPI-EMS-LR
18. FIO-ESM	39. MPI-ESM-MR
19. GFDL-CM2.1	40. MRI-CGCM3
20. GFDL-CM3	41. NorESM1-M
21. GFDL-ESM2G	42. NorESM1-ME

(b) for GISS, 0.1002 °C is first added to each value of ΔT; 0.1002 °C is the difference between the mean value of GISS ΔT during 1961–1990 relative to 1951–1980. After this initial addition, the GISS data represent ΔT relative to 1961–1990. A second addition of 0.3134 °C then occurs, to place the data on the 1850–1900 baseline;

c) for NCEI, 0.1202 °C is first subtracted from each value of ΔT; 0.1202 °C is the difference between the mean value of NCEI ΔT during 1961–1990 relative to 1901–2000. After this initial addition, the NCEI data represent ΔT relative to 1961–1990. A second addition of 0.3134 °C then occurs, to place the data on the 1850–1900 baseline.

The GCM lines in the figure are based on analysis of all of the r*i1p1 files present on the CMIP5 archive as of early summer 2016. The 42 GCMs considered are given in Table 2.2. According to the CMIP5 nomenclature, "r" refers to realization, "i" refers to initialization method, and "p" refers to physics version, and "*" is notation for any integer. The integer that appears after the "r" in the GCM output file name is used to distinguish members of an ensemble, or realization, generated by initializing a set of GCM runs with different but equally realistic initial conditions; the "i" in the file name refers to a different method of initializing the GCM simulation; and, the "p" denotes perturbed GCM model physics. The string i1p1 appears in the vast majority of the archived files.

Table 2.3 AAWR from GCM RCP 4.5 simulations in the CMIP5 archive

CMIP5 GCM	Modeling Center	Ensemble run	GCM-AAWR (°C/dec)	
			LIN	REG
ACCESS1.0	Bureau of Meteorology, Australia	r1i1p1	0.248	**0.230**
ACCESS1.3		r1i1p1	0.234	**0.206**
		Ctr Avg	0.241	0.218
BCC-CSM1.1	Beijing Climate Center, China Meteorological Administration	r1i1p1	0.259	**0.253**
BCC-CSM1.1(m)		r1i1p1	0.286	**0.278**
		Ctr Avg	0.273	0.265
BNU-ESM	College of Global Change and Earth System Science, Beijing Normal University, China	r1i1p1	0.320	**0.301**
CCSM4	National Center for Atmospheric Research (NCAR), United States	r1i1p1	0.284	0.280
		r2i1p1	0.255	0.247
		r3i1p1	0.226	0.225
		r4i1p1	0.214	0.204
		r5i1p1	0.283	0.252
		r6i1p1	0.234	0.223
		Mod Avg	0.249	**0.238**
CESM1(BGC)	Community Earth System	r1i1p1	0.249	**0.223**
CESM1(CAM5)	Model Contributors, NCAR, United States	r1i1p1	0.198	0.179
		r2i1p1	0.193	0.184
		r3i1p1	0.243	0.230
		Mod Avg	0.211	**0.198**
		Ctr Avg	0.232	0.204
CMCC-CM	Centro Euro–Mediterraneo per I Cambiamenti Climatici, France	r1i1p1	0.228	**0.235**
CMCC-CMS		r1i1p1	0.227	**0.250**
CNRM-CM5		r1i1p1	0.242	**0.221**
		Ctr Avg	0.232	0.236
CSIRO-Mk3.6.0	Commonwealth Scientific and Industrial Research Organization, Australia	r1i1p1	0.172	**0.170**

CanCM4	Canadian Centre for Climate Modelling and Analysis		
		r1i1p1	0.243	0.226
		r2i1p1	0.267	0.260
		r3i1p1	0.230	0.219
		r4i1p1	0.289	0.279
		r5i1p1	0.226	0.220
		r6i1p1	0.228	0.220
		r7i1p1	0.278	0.249
		r8i1p1	0.265	0.252
		r9i1p1	0.214	0.204
		r10i1p1	0.195	0.191
		Mod Avg	0.244	**0.232**
CanESM2		r1i1p1	0.321	0.286
		r2i1p1	0.334	0.315
		r3i1p1	0.307	0.295
		r4i1p1	0.331	0.302
		r5i1p1	0.326	0.308
		Mod Avg	0.324	**0.301**
		Ctr Avg	0.270	0.255
EC-EARTH	EC-EARTH consortium (numerous national weather services and universities, from 11 countries in Europe, participate in this effort)	r1i1p1	0.220	0.209
		r2i1p1	0.187	0.178
		r5i1p1	0.210	0.197
		r6i1p1	0.157	0.146
		r8i1p1	0.204	0.203
		r9i1p1	0.186	0.181
		r12i1p1	0.155	0.149
		r13i1p1	0.233	0.233
		r14i1p1	0.188	0.160
		Mod Avg	0.193	**0.184**

(continued)

Table 2.3 (continued)

CMIP5 GCM	Modeling Center	Ensemble run	GCM-AAWR (°C/dec)	
			LIN	REG
FGOALS-g2	Institute of Atmos. Physics, Chinese Academy of Sciences	r1i1p1	0.179	**0.185**
FIO-ESM	First Institute of Oceanography, State Oceanic Administration, China	r1i1p1	0.188	0.192
		r2i1p1	0.184	0.187
		r3i1p1	0.203	0.207
		Mod Avg	0.191	**0.195**
GFDL-CM2.1	NOAA Geophysical Fluid Dynamics Laboratory, United States	r1i1p1	0.261	0.250
		r2i1p1	0.319	0.319
		r3i1p1	0.297	0.266
		r4i1p1	0.294	0.262
		r5i1p1	0.301	0.287
		r6i1p1	0.197	0.203
		r7i1p1	0.253	0.226
		r8i1p1	0.274	0.278
		r9i1p1	0.202	0.194
		r10i1p1	0.263	0.245
		Mod Avg	0.266	**0.253**
GFDL-CM3		r1i1p1	0.270	**0.257**
GFDL-ESM2G		r1i1p1	0.275	**0.253**
GFDL-ESM2M		r1i1p1	0.204	**0.183**
		Ctr Avg	0.262	0.248

GISS-E2-H	NASA Goddard Institute for Space Studies, United States	r1i1p1	0.192	0.174
		r2i1p1	0.216	0.194
		r3i1p1	0.192	0.186
		r4i1p1	0.207	0.192
		r5i1p1	0.178	0.171
		Mod Avg	0.197	**0.183**
GISS-E2-H-CC		r1i1p1	0.222	**0.214**
GISS-E2-R		r1i1p1	0.185	0.169
		r2i1p1	0.189	0.177
		r3i1p1	0.193	0.181
		r4i1p1	0.169	0.171
		r5i1p1	0.141	0.136
		r6i1p1	0.229	0.204
		Mod Avg	0.184	**0.173**
GISS-E2-R-CC		r1i1p1	0.200	**0.191**
		Ctr Avg	0.193	0.182

(continued)

Table 2.3 (continued)

CMIP5 GCM	Modeling Center	Ensemble run	GCM-AAWR (°C/dec)	
			LIN	REG
HadCM3	Met Office Hadley Centre, United Kingdom. Additional HadGEM2-ES realizations were contributed by Instituto Nacional de Pesquisas Espaciais, Brazil	r1i1p1	0.235	0.236
		r2i1p1	0.200	0.171
		r3i1p1	0.250	0.230
		r4i1p1	0.208	0.192
		r5i1p1	0.297	0.271
		r6i1p1	0.192	0.195
		r7i1p1	0.258	0.236
		r8i1p1	0.257	0.214
		r9i1p1	0.23	0.217
		r10i1p1	0.233	0.215
		Mod Avg	0.236	**0.218**
HadGEM2-CC		r1i1p1	0.184	**0.183**
HadGEM2-ES		r1i1p1	0.289	0.277
		r2i1p1	0.204	0.195
		r3i1p1	0.185	0.177
		r4i1p1	0.274	0.233
		Mod Avg	0.238	**0.221**
		Ctr Avg	0.233	0.216
INM-CM4	Institute for Numerical Mathematics, Russian Academy of Sciences	r1i1p1	0.100	**0.098**
IPSL-CM5A-LR	Institut Pierre-Simon Laplace, France	r1i1p1	0.323	0.317
		r2i1p1	0.297	0.294
		r3i1p1	0.216	0.220
		r4i1p1	0.256	0.248
		Mod Avg	0.273	**0.270**
IPSL-CM5A-MR		r1i1p1	0.253	**0.235**
IPSL-CM5B-LR		r1i1p1	0.122	**0.122**
		Ctr Avg	0.244	0.239

MIROC-ESM	Japan Agency for Marine-Earth Science and Technology, Atmosphere and Ocean Research Institute (Univ. of Tokyo), and National Institute for Environmental Studies	r1i1p1	0.177	**0.172**
MIROC-ESM-CHEM		r1i1p1	0.170	**0.156**
		Ctr Avg	0.174	0.164
MIROC4h	Atmosphere and Ocean Research Institute (Univ. of Tokyo), National Institute for Environmental Studies, and Japan Agency for Marine-Earth Science and Technology	r1i1p1	0.252	0.251
		r2i1p1	0.300	0.282
		r3i1p1	0.317	0.299
		Mod Avg	0.290	**0.277**
MIROC5		r1i1p1	0.278	0.273
		r2i1p1	0.187	0.154
		r3i1p1	0.287	0.256
		Mod Avg	0.251	**0.228**
		Ctr Avg	0.270	0.252
MPI-ESM-LR	Max-Planck-Institut für Meteorologie (Max Planck Institute for Meteorology), Germany	r1i1p1	0.161	0.144
		r2i1p1	0.248	0.224
		r3i1p1	0.212	0.205
		Mod Avg	0.207	**0.191**
MPI-ESM-MR		r1i1p1	0.272	0.256
		r2i1p1	0.199	0.184
		r3i1p1	0.239	0.225
		Mod Avg	0.237	**0.222**
		Ctr Avg	0.222	0.206
MRI-CGCM3	Meteorological Research Institute, Japan	r1i1p1	0.089	**0.075**
NorEMS1-M	Norwegian Climate Centre	r1i1p1	0.156	**0.157**
NorEMS1-ME		r1i1p1	0.180	**0.172**
		Ctr Avg	0.168	0.164

(continued)

For a GCM to have been used, a historical file had to have been submitted to the CMIP5 archive. The historical files contain output of gridded surface temperatures, generally for the 1850–2005 time period. Global mean surface temperature is computed, using cosine latitude weighting. Next, an offset such that GMST from the historical run of each GCM can be placed onto a 1961–1990 baseline is found and recorded. This offset is applied to all of the r*i1p1 files from the future runs of the specific GCM, which generally cover the 2006–2100 time period. All GCM time series are then placed onto the 1850v1900 baseline by adding 0.3134 °C to each value of ΔT. All of the GCMs except CCM-CESM listed in Table 2.2 submitted future runs for RCP 4.5 to the CMIP5 archive; a single line for each of the other 41 models appears in Fig. 2.3a. For RCP 8.5, all of the GCMs except CanCM4, GFDL-CM2.1, HadCM3, and MIROC4h submitted output for RCP 8.5 to the CMIP5 archive; a single line for each of the other 38 models appears in Fig. 2.3b. Information about the Modeling Center and Institution for these models is provided in our Table 2.3 below, for models that submitted results for RCP 4.5, and on the web at http://cmip-pcmdi.llnl.gov/cmip5/docs/CMIP5_modeling_groups.pdf.

Figure 2.3 also contains a green trapezoid and vertical bar. The coordinates of the trapezoid are (2016, 0.722 °C), (2016, 1.092 °C), (2035, 0.877 °C) and (2035, 1.710 °C) and the coordinates of the vertical bar are (2026, 0.89 °C) and (2026, 1.29 °C). Anyone concerned about the veracity of Fig. 2.3 is urged to have a look at Fig. 11.25 of IPCC (2013). The right hand side of Fig. 11.25b includes an axis labeled "Relative to 1850–1900". Our Fig. 2.3 visually matches Fig. 11.25 of IPCC (2013) to a very high level of quantitative detail.

Figures 2.4 and **2.5** compare ΔT relative to the 1850–1900 baseline from CRU to values of ΔT found using the Empirical Model of Global Climate. Values of model output parameters λ, κ, ECS, and AAWR are all recorded in Fig. 2.4. The simulation in Fig. 2.4 was found upon setting the regression coefficients C_4, C_5, and C_6 in Eq. 2.2 to zero. The simulation in Fig. 2.5 made full use of all regression coefficients. The comparison of modeled and measured OHC that corresponds to the simulation shown in Fig. 2.5 is nearly identical to the bottom panel of Fig. 2.4, and hence has been omitted. The same value of κ was found for both of these simulations. The bottom two rungs of Fig. 2.5 show the contribution to modeled ΔT from AMOC, PDO, and IO; the slope of the AMOC contribution over 1979–2010 is also recorded. The top rung of each ladder plot also records the goodness of fit parameter χ^2 (Eq. 2.7) for the two simulations. Finally, the top two rungs of each ladder plot are labeled "ΔT from preindustrial" whereas the other rungs have labels of ΔT. The label ΔT is used for the lower rungs for compactness of notation.

Figure 2.6 shows time series for ΔRF of six classes of anthropogenic, tropospheric aerosols: four that tend to cool climate (sulfate, organic carbon from combustion of fossil fuels, dust, and nitrate) and two that warm (black carbon from combustion of fossil fuels and biomass burning, and organic carbon from biomass burning). Estimates of direct ΔRF from all but sulfate originate from values of direct radiative forcing of climate obtained from file:

RCP45_MIDYEAR_RADFORCING.DAT provided by PICR at: http://www.pik-potsdam.de/~mmalte/rcps/data

We have modified the PICR value for direct radiative forcing of sulfate, using data from Stern (2006a, b), and Smith et al. (2011), as described in our methods paper (Canty et al. 2013), because the modified time series is deemed to be more accurate than the RCP value, which was based on projections of sulfate emission reductions conducted prior to the publication of Smith et al. (2011).

The estimates of direct ΔRF from the various aerosol types are then combined into two time series: one for the aerosols that cool, the other for the aerosols that heat. Next, these two time series are multiplied by scaling parameters that represent the aerosol indirect effect[33] for aerosols that cool and for aerosols that warm. These are the six curves shown using colors that correspond to aerosol type. The total direct ΔRF of aerosols that warm, and aerosols that cool, are shown by the red and blue lines, respectively. The line labeled Net is the sum of the total warming and total cooling term, and reflects the time series of Aerosol ΔRF_i input to the EM-GC (Eq. 2.2). Finally, the black open square marks $AerRF_{2011} = -0.9$ W m^{-2} along the Net time series, which is the best estimate of total ΔRF due to anthropogenic tropospheric aerosols given by IPCC (2013).

Canty et al. (2013) relied on scaling parameters that were tied to numerical estimates of upper and lower limits of the aerosol indirect effect given by IPCC (2007) (their Fig. 4). Figure 2.21 is our new scaling parameter "road map", updated to reflect estimates of the aerosol indirect effect by IPCC (2013). The set of scaling parameters used in Fig. 2.6 are given by the intersection of "Middle Road" with the AerRF2011 = -0.9 W m^{-2} line in Fig. 2.21: i.e., $\alpha_{HEAT} = 2.19$ and $\alpha_{COOL} = 2.43$. Further details of our approach for assessing a wide range of aerosol ΔRF scenarios in a manner consistent with both CMIP5 and IPCC is given in Canty et al. (2013).

Figure 2.7 shows time series of Aerosol ΔRF_i found using scaling parameters α_{HEAT} and α_{COOL}, combined with estimates of direct ΔRF of climate found as described above, for five values of $AerRF_{2011}$: -0.1, -0.4, -0.9, -1.5, and -1.9 W m^{-2} (open squares). The highest and lowest values of $AerRF_{2011}$ are the upper and lower limits of the possible range, the second highest and second lowest values are the limits of the likely range, and the middle value is the best estimate, all from IPCC (2013). Three curves are shown for each value of $AerRF_{2011}$: the solid curve uses values for scaling parameters α_{HEAT} and α_{COOL} along the Middle Road of Fig. 2.21, whereas the other lines use parameters along the High and Low Roads.

Figure 2.8 shows time series of ocean heat content for the upper 700 m of earth's oceans from six sources, as indicated. The data have all been normalized to a common value of zero, at the start of 1993. This normalization is done for visual convenience; the EM-GC model simulates OHE, which is the time rate of change of OHC. The time rate of change is the slope of each dataset, which is unaltered upon application of an offset. The data sources are:

Balmaseda et al. (2013): http://www.cgd.ucar.edu/cas/catalog/ocean/OHC700m.tar.gz
Church et al. (2011): http://www.cmar.csiro.au/sealevel/TSL_OHC_20110926.html

---

[33] The aerosol indirect effect is scientific nomenclature for changes in the radiative forcing of climate due to modifications to clouds caused by anthropogenic aerosols.

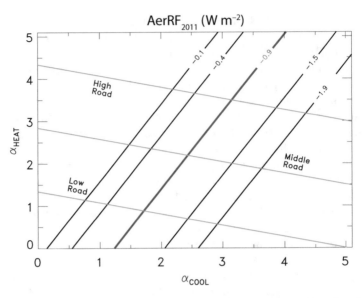

Fig. 2.21 Aerosol indirect effect scaling parameters. The *black lines* show values of total ΔRF of climate in year 2011 (AerRF$_{2011}$), relative to pre-industrial baseline, due to anthropogenic aerosols, as a function of the parameter used to multiply the total direct ΔRF of climate from all aerosols that cool (α_{COOL}) and the parameter used to multiply the total direct ΔRF of climate from all aerosols that heat (α_{HEAT}). Parameters α_{COOL} and α_{HEAT} represent the effect of aerosols on the occurrence, distribution, and properties of clouds: the so-called aerosol indirect effects. The red line shows the most likely value of AerRF$_{2011}$. − 0.9 W m^{-2}, from IPCC (2013). The *black lines* represent the IPCC (2013) upper and lower limits of the likely range (−0.4 and −1.5 W m^{-2}) and the upper and lower limits of the possible range for AerRF$_{2011}$ (−0.1 and −1.9 W m^{-2}). This figure is included to indicate that various combinations of α_{COOL} and α_{HEAT} can be used to find a particular value of AerRF$_{2011}$. The combination of parameters along the line marked Middle Road is the most likely combination of parameters, based on detailed examination of various tables given in Chap. 7 of IPCC (2013). The high road and low road represent the ranges of plausible values of scaling parameters, again based on our analysis of (IPCC 2013). Further details about this approach for representing the aerosol indirect effect in the EM-GC are given in our methods paper (Canty et al. 2013)

Giese et al. (2011): http://dsrs.atmos.umd.edu/DATA/soda_hc2_700.nc

Gouretski and Reseghetti (2010): http://www1.ncdc.noaa.gov/pub/data/cmb/bams-sotc/2009/global-data-sets/OHC_viktor.txt

Ishii and Kimoto (2009): http://www1.ncdc.noaa.gov/pub/data/cmb/bams-sotc/2009/global-data-sets/OHC_ishii.txt

Levitus et al. (2012): http://data.nodc.noaa.gov/woa/DATA_ANALYSIS/3M_HEAT_CONTENT/DATA/basin/yearly/h22-w0-700m.dat

As explained in the text, values of OHC shown in Fig. 1.8 are multiplied by 1/0.7 = 1.42 prior to being used in the EM-GC, to represent the estimate that 70 % of the rise in OHC occurs in the upper 700 m of the world's oceans (Sect. 5.2.2.1 of IPCC 2007).

Figure 2.11 shows twelve estimates of ECS. The six to the left are previously published values and the six to the right are values found using our EM-GC. Here, numerical estimates of the circle (best estimate), range, and brief description are given.

The ECS value from IPCC (2007) of 3.3 (2.1, 4.4) °C, given in Box 10.2, is based on GCMs that contributed to this report. Here, 2.1 and 4.4 °C are the lower and upper limits of ECS, based on <5 % and >95 % probabilities (i.e., 95 % confidence interval), respectively, as explained in Box TS.1 of IPCC (2007). The entry from Shindell et al. (2013) of 4.0 (2.4, 4.7) °C represents the mean and ranges (lower and upper limit) of the value of ECS from eight GCMs given in Fig. 22 of their paper. The value from IPCC (2013) of 3.2 (1.9, 4.5) °C is from Table 9.5 that provides ECS for 23 GCMs; here, the limits represents 90 % confidence intervals.

The ECS value from Schwartz (2012) of 2.23 (1.06, 3.40) °C represents the mean and standard deviation of the nine determinations given in Table 2.2 of this paper. The value from Otto et al. (2013) of 2.0 (1.2, 3.9) °C is the most likely value and 95 % confidence interval uncertainty for the first decade of this century. Finally, the ECS from Masters (2014) of 1.98 (1.19, 5.15) °C is the most likely value and 90 % confidence interval from an analysis that covered the past 50 years.

For the EM-GC based estimates of ECS, the error bars represent the range of uncertainty for consideration of the IPCC (2013) expert judgement of the upper limits of the full possible range of $AerRF_{2011}$ (i.e., -0.1 and -1.9 W m^{-2}) and each circle show the value of ECS found for $AerRF_{2011}$ equal to -0.5 W m^{-2}, the IPCC best estimate.

Figure 2.12 shows Attributable Anthropogenic Warming Rate (AAWR) as a function of ΔRF due to aerosols. As for many of our analyses, results are shown for five values of $AerRF_{2011}$:-0.1. -0.4, -0.9, -1.5, and -1.9 W m^{-2} :which define the possible range, the likely range, and best estimate of $AERRF_{2011}$ according to IPCC (2013). For each value of $AerRF_{2011}$, model runs are conducted for the three determinations of Aerosol ΔRF shown in Fig. 2.7a. The circle represents the mean of these three runs; the error bars represent the maximum and minimum values. Precise determination of AAWR does depend on knowledge of how aerosol ΔRF has varied over the time period of interest; uncertainty in the shape of aerosol ΔRF over 1979–2010 exerts considerable influence on AAWR.

Figure 2.13 shows AAWR from numerous EM-GC simulations, as detailed in the caption, and AAWR found from the 41 GCMs that submitted RCP 4.5 future runs to the CMIP5 archive. Here, a detailed explanation is provided for the determination of GCM-based AAWR.

The estimate of AAWR from GCMs is based on analysis of 112 runs of 41 GCMs, from 21 modeling centers, submitted to the CMIP5 archive. AAWR has been computed for each run using two methods: regression (REG) and linear fit (LIN). Table 2.3 details the 112 determinations of AAWR, from each method, organized first by the name of each GCM, then by modeling center. As noted earlier, we use all of the r*i1p1 runs in the CMIP5 archive that cover both the historical time period (these runs generally stop at year 2005) and the future for RCP 4.5 forcing (these runs generally start at 2006). According to CMIP5 nomenclature, "r" refers to different realizations of an ensemble simulation, all of which are initialized with different but equality realistic initial conditions; "i" refers to a completely different method for initializing a particular GCM simulation; and, "p" denotes some perturbation to GCM model physics. The string r*i1p1 appears in the vast majority of CMIP5 files; examination of the 112 r*i1p1 runs provides a robust examination of GCM output.

The first method used to extract AAWR from each GCM run, REG, involves examination of de-seasonalized, globally averaged, monthly mean values of ΔT from each run, from 1950 to 2010. Archived model output from the historical and the future run files has been combined. Both the historical and future runs were designed to use realistic variations of total solar irradiance (TSI) and stratospheric optical depth (SOD), the climate relevant proxy for major volcanic eruptions. First, regression coefficients for TSI, SOD, and ΔT^{HUMAN} are found. For this first step, observations of TSI and SOD are used in the analysis, and ΔT^{HUMAN} is approximated as a linear function. The regression coefficient for TSI is saved. A second regression is conducted using ΔT from the GCM, for the 1979–2010 time period. For the second regression, the saved value for the TSI coefficient is imposed, leading to new values for the coefficients that modify SOD and ΔT^{HUMAN}. A two step method is needed to properly determine the TSI and SOD coefficients, because the two major volcanic eruptions that took place over the period of interest, El Chichón and Mount Pinatubo, occurred at similar phases of the 11 year solar cycle. The initial regression starts in 1950 to allow coverage of enough solar cycles for extraction of the influence of solar variability on GCM-based ΔT to be found, and also because ΔT^{HUMAN} over 1950–2010 found using EM-GC (i.e., Human Rung on the Figs. 2.4, 2.5, 2.9, and 2.10 ladder plots) is nearly linear over this 60 year time frame. The value of AAWR using REG is the slope of ΔT^{HUMAN}, recorded for each of the 112 GCM runs in Table 2.3.

The second method used to extract AAWR from each GCM run, LIN, involves analysis of global, annual average values of ΔT from the various GCM runs. As noted above, these GCM runs were designed to simulate the short-term cooling caused by volcanic eruptions, such as El Chichón and Mount Pinatubo. The volcanic imprint from most of the GCM runs is obvious upon visual inspection: archived ΔT tends to be smaller than neighboring years in 1982, 1983, 1991, and 1992. For LIN, we find the slope of global annual average ΔT from each GCM run using linear regression, excluding archived output for the four years noted in the prior sentence. Values of AAWR found using LIN are also recorded for each of the 112 GCM runs in Table 2.3.

We are confident AAWR has been properly extracted from the archived GCM output. Neither of our determinations attempt to discern the influence on GCM-based ΔT of natural variations such as ENSO, PDO, or AMOC. While the CMIP5 GCMs represent ENSO with some fidelity (Bellenger et al. 2014), and changes in heat storage within the Pacific ocean simulated by GCMs has been linked to variability in ΔT on decadal time scales (Meehl et al. 2011), these effects should appear as noise that is averaged out of the resulting signal, since our estimates of AAWR are based on analysis of 112 archived GCM runs. While GCMs might indeed have internally generated ENSO events or fluctuations in ocean heat storage that affect

ΔT, the years in which these modeled events occur will bear no relation to the years these events occur in the real world (or in other models). A detailed examination of model output from four leading research centers finds little impact on ΔT of variations in the strength of AMOC within GCMs (Kavvada et al. 2013). Conversely, accurate timing of natural variations of ΔT due to solar irradiance and volcanoes is imposed on GCMs, via request that the GCMs use actual variations in TSI and SOD derived from data.

Statistical analysis supports the contention that the representation of GCM-based AAWR in Fig. 2.3 is accurate. The 112 values of AAWR in Table 2.3 found using REG compared to the 112 values found using LIN result in a correlation coefficient (r^2) of 0.953 and a ratio of 1.057 ± 0.106, with AAWR LIN tending to exceed AAWR REG by 5.7 %. Consideration of the values of AAWR associated with the 41 GCMs yields $r^2 = 0.964$ and ratio of 1.051 ± 0.101; again AAWR LIN is slightly larger than AAWR REG. Finally, analysis of AAWR from the 21 modeling centers yields $r^2 = 0.977$ and ratio = 1.052 ± 0.103. Values of AAWR found using REG and LIN agree to within 5 % with a variance of 10 %. We conclude our determination of GCM-based AAWR is accurate to ±10 %, which is much smaller than the difference between the GCM-based value of AAWR and that found using the EM-GC framework shown in Fig. 2.13.

The box and whisker (BW) symbol in Fig. 2.13 is based on AAWR found using the regression method (REG), for all 41 GCMs that submitted RCP 4.5 output to the CMIP5 archive. If a model submitted multiple runs, the resulting AAWR values are averaged, leading to a single value of AAWR for each GCM.[34] The 41 values of AAWR upon which the BW plot is based are bold-faced on Table 2.3. The resulting BW symbol for the values of AAWR found using the linear fit (LIN) method, for the 41 GCMs in Table 2.3, is quite similar to the BW symbol shown in Fig. 2.13. The primary difference is a higher median value for the LIN determination: the 25th, 75th, minimum, and maximum values are quite similar to those of the REG method. Finally, BW symbols for AAWR based on either the 112 runs or the 21 modeling centers, found using either LIN or REG, look quite similar to the GCM representation in Fig. 2.13.

References

Ammann CM, Meehl GA, Washington WM, Zender CS (2003) A monthly and latitudinally varying volcanic forcing dataset in simulations of 20th century climate. Geophys Res Lett 30(12):1657. doi:10.1029/2003GL016875

Andrews T, Gregory JM, Webb MJ, Taylor KE (2012) Forcing, feedbacks and climate sensitivity in CMIP5 coupled atmosphere-ocean climate models. Geophys Res Lett 39(9). doi:10.1029/2012gl051607

---

[34] Nearly identical values of AAWR are found if, rather than averaging the multiple determinations, the time series of ΔT from each GCM are averaged, and a single value of AAWR is found from the resulting, averaged time series.

Andronova NG, Schlesinger ME (2000) Causes of global temperature changes during the 19th and 20th centuries. Geophys Res Lett 27(14):2137–2140. doi:10.1029/2000GL006109

Ballantyne AP, Alden CB, Miller JB, Tans PP, White JWC (2012) Increase in observed net carbon dioxide uptake by land and oceans during the past 50 years. Nature 488(7409):70–72

Balmaseda MA, Trenberth KE, Källén E (2013) Distinctive climate signals in reanalysis of global ocean heat content. Geophys Res Lett 40(9):1754–1759. doi:10.1002/grl.50382

Bellenger H, Guilyardi E, Leloup J, Lengaigne M, Vialard J (2014) ENSO representation in climate models: from CMIP3 to CMIP5. Clim Dyn 42(7):1999–2018. doi:10.1007/s00382-013-1783-z

Bond TC, Doherty SJ, Fahey DW, Forster PM, Berntsen T, DeAngelo BJ, Flanner MG, Ghan S, Kärcher B, Koch D, Kinne S, Kondo Y, Quinn PK, Sarofim MC, Schultz MG, Schulz M, Venkataraman C, Zhang H, Zhang S, Bellouin N, Guttikunda SK, Hopke PK, Jacobson MZ, Kaiser JW, Klimont Z, Lohmann U, Schwarz JP, Shindell D, Storelvmo T, Warren SG, Zender CS (2013) Bounding the role of black carbon in the climate system: a scientific assessment. J Geophys Res Atmos 118(11):5380–5552. doi:10.1002/jgrd.50171

Bony S, Colman R, Kattsov VM, Allan RP, Bretherton CS, Dufresne J-L, Hall A, Hallegatte S, Holland MM, Ingram W, Randall DA, Doden BJ, Tselioudis G, Webb MJ (2006) How well do we understand and evaluate climate change feedback processes? J Clim 19:3445–3482. doi:10.1029/2005GL023851

Cai W, van Rensch P, Cowan T, Hendon HH (2011) Teleconnection pathways of ENSO and the IOD and the mechanisms for impacts on Australian rainfall. J Clim 24(15):3910–3923. doi:10.1175/2011JCLI4129.1

Canty T, Mascioli NR, Smarte MD, Salawitch RJ (2013) An empirical model of global climate—Part 1: A critical evaluation of volcanic cooling. Atmos Chem Phys 13(8):3997–4031. doi:10.5194/acp-13-3997-2013

Carton JA, Giese BS (2008) A reanalysis of ocean climate using simple ocean data assimilation (SODA). Mon Weather Rev 136(8):2999–3017. doi:10.1175/2007MWR1978.1

Carton JA, Santorelli A (2008) Global decadal upper-ocean heat content as viewed in nine analyses. J Clim 21(22):6015–6035. doi:10.1175/2008jcli2489.1

Chavez FP, Ryan J, Lluch-Cota SE, Niquen CM (2003) From anchovies to sardines and back: multidecadal change in the Pacific Ocean. Science 299(5604):217–221. doi:10.1126/science.1075880

Church JA, White NJ, Konikow LF, Domingues CM, Cogley JG, Rignot E, Gregory JM, van den Broeke MR, Monaghan AJ, Velicogna I (2011) Revisiting the Earth's sea-level and energy budgets from 1961 to 2008. Geophys Res Lett 38(18):L18601. doi:10.1029/2011GL048794

Chylek P, Klett JD, Lesins G, Dubey MK, Hengartner N (2014) The Atlantic Multidecadal Oscillation as a dominant factor of oceanic influence on climate. Geophys Res Lett 41. doi:10.1002/2014GL059274

Coddington O, Lean JL, Pilewskie P, Snow M, Lindholm D (2016) A solar irradiance climate data record. Bull Am Meteorol Soc. doi:10.1175/BAMS-D-14-00265.1

Cowtan K, Way RG (2014) Coverage bias in the HadCRUT4 temperature series and its impact on recent temperature trends. Q J R Meteorol Soc 140(683):1935–1944. doi:10.1002/qj.2297

DelSole T, Tippett MK, Shukla J (2011) A significant component of unforced multidecadal variability in the recent acceleration of global warming. J Clim 24(3):909–926. doi:10.1175/2010jcli3659.1

Dessler AE (2010) A determination of the cloud feedback from climate variations over the past decade. Science 330(6010):1523–1527. doi:10.1126/science.1192546

Dlugokencky EJ, Bruhwiler L, White JWC, Emmons LK, Novelli PC, Montzka SA, Masarie KA, Lang PM, Crotwell AM, Miller JB, Gatti LV (2009) Observational constraints on recent increases in the atmospheric CH$_4$ burden. Geophys Res Lett 36(18):L18803. doi:10.1029/2009GL039780

Douglass DH, Knox RS (2005) Climate forcing by the volcanic eruption of Mount Pinatubo. Geophys Res Lett 32(5):L05710. doi:10.1029/2004GL022119

Duchez A, Hirschi JJ-M, Cunningham SA, Blaker AT, Bryden HL, de Cuevas B, Atkinson CP, McCarthy GD, Frajka-Williams E, Rayner D, Smeed D, Mizielinski MS (2014) A new index for the Atlantic Meridional Overturning Circulation at 26°N. J Clim 27(17):6439–6455. doi:10.1175/JCLI-D-13-00052.1

England MH, McGregor S, Spence P, Meehl GA, Timmermann A, Cai W, Gupta AS, McPhaden MJ, Purich A, Santoso A (2014) Recent intensification of wind-driven circulation in the Pacific and the ongoing warming hiatus. Nat Clim Chang 4(3):222–227. doi:10.1038/nclimate2106

Etheridge DM, Steele LP, Francey RJ, Langenfelds RL (1998) Atmospheric methane between 1000 A.D. and present: Evidence of anthropogenic emissions and climatic variability. J Geophys Res Atmos 103(D13):15979–15993. doi:10.1029/98JD00923

Foster G, Rahmstorf S (2011) Global temperature evolution 1979–2010. Environ Res Lett 6(4):044022. doi:10.1088/1748-9326/6/4/044022

Fromm M, Kablick G, Nedoluha G, Carboni E, Grainger R, Campbell J, Lewis J (2014) Correcting the record of volcanic stratospheric aerosol impact: Nabro and Sarychev Peak. J Geophys Res Atmos 119(17):10,343–310,364. doi:10.1002/2014JD021507

Giese BS, Chepurin GA, Carton JA, Boyer TP, Seidel HF (2011) Impact of bathythermograph temperature bias models on an ocean reanalysis. J Clim 24(1):84–93. doi:10.1175/2010jcli3534.1

Gillett NP, Arora VK, Matthews D, Allen MR (2013) Constraining the ratio of global warming to cumulative CO_2 emissions using CMIP5 simulations. J Clim 26(18):6844–6858. doi:10.1175/JCLI-D-12-00476.1

Gouretski V, Reseghetti F (2010) On depth and temperature biases in bathythermograph data: development of a new correction scheme based on analysis of a global ocean database. Deep-Sea Res I Oceanogr Res Pap 57(6):812–833. doi:10.1016/j.dsr.2010.03.011

Hansen JE, Ruedy R, Sato M, Lo K (2010) Global surface temperature change. Rev Geophys 48(4). doi:10.1029/2010rg000345

Hansen JE, Sato M, Kharecha P, von Schuckmann K (2011) Earth's energy imbalance and implications. Atmos Chem Phys 11(24):13421–13449. doi:10.5194/acp-11-13421-2011

Houghton JT (2015) Global warming: the complete briefing, 5th edn. Cambridge University Press, Cambridge

IPCC (2001) Climate change 2001: the scientific basis. Contribution of working group I to the third assessment report of the intergovernmental panel on climate change. Cambridge, UK and New York, NY, USA

IPCC (2007) Climate change 2007: the physical science basis. Contribution of working group I to the fourth assessment report of the intergovernmental panel on climate change. Cambridge, UK and New York, NY, USA

IPCC (2013) Climate change 2013: The physical science basis. Contribution of working group I to the fifth assessment report of the intergovernmental panel on climate change. Cambridge, UK and New York, NY, USA

Ishii M, Kimoto M (2009) Reevaluation of historical ocean heat content variations with time-varying XBT and MBT depth bias corrections. J Oceanogr 65(3):287–299, doi: 10.1007/s10872-009-0027-7

Jones PD, Lister DH, Osborn TJ, Harpham C, Salmon M, Morice CP (2012) Hemispheric and large-scale land-surface air temperature variations: an extensive revision and an update to 2010. J Geophys Res 117(D5):D05127. doi:10.1029/2011jd017139

Kahn RA (2012) Reducing the uncertainties in direct aerosol radiative forcing. Surv Geophys 33(3):701–721. doi:10.1007/s10712-011-9153-z

Karl TR, Arguez A, Huang B, Lawrimore JH, McMahon JR, Menne MJ, Peterson TC, Vose RS, Zhang H-M (2015) Possible artifacts of data biases in the recent global surface warming hiatus. Science 348(6242):1469–1472. doi:10.1126/science.aaa5632

Kavvada A, Ruiz-Barradas A, Nigam S (2013) AMO's structure and climate footprint in observations and IPCC AR5 climate simulations. Clim Dyn 41(5-6):1345–1364. doi:10.1007/s00382-013-1712-1

Kennedy JJ, Rayner NA, Smith RO, Parker DE, Saunby M (2011a) Reassessing biases and other uncertainties in sea-surface temperature observations measured in situ since 1850, Part 1: Measurement and sampling uncertainties. J Geophys Res 116:D14103. doi:10.1029/201 0JD015218

Kennedy JJ, Rayner NA, Smith RO, Parker DE, Saunby M (2011b) Reassessing biases and other uncertainties in sea-surface temperature observations measured in situ since 1850, Part 2: Biases and homogenisation. J Geophys Res 116:D14104. doi:10.1029/2010JD015220

Kiehl JT (2007) Twentieth century climate model response and climate sensitivity. Geophys Res Lett 34(22):L22710. doi:10.1029/2007GL031383

Knight JR, Allan RJ, Folland CK, Vellinga M, Mann ME (2005) A signature of persistent natural thermohaline circulation cycles in observed climate. Geophys Res Lett 32(20):L20708. doi:10 .1029/2005GL024233

Kosaka Y, Xie SP (2013) Recent global-warming hiatus tied to equatorial Pacific surface cooling. Nature 501(7467):403–407. doi:10.1038/nature12534

Lean JL (2000) Evolution of the Sun's spectral irradiance since the maunder minimum. Geophys Res Lett 27(16):2425–2428. doi:10.1029/2000GL000043

Lean JL, Rind DH (2008) How natural and anthropogenic influences alter global and regional surface temperatures: 1889 to 2006. Geophys Res Lett 35(18):L18701. doi:10.1029/200 8GL034864

Lean JL, Rind DH (2009) How will Earth's surface temperature change in future decades? Geophys Res Lett 36(15):L15708. doi:10.1029/2009GL038932

Levitus S, Antonov JI, Boyer TP, Baranova OK, Garcia HE, Locarnini RA, Mishonov AV, Reagan JR, Seidov D, Yarosh ES, Zweng MM (2012) World ocean heat content and thermosteric sea level change (0–2000 m), 1955–2010. Geophys Res Lett 39(10):L10603. doi:10.1029/201 2GL051106

Lynas M (2008) Six degrees: our future on a hotter planet. National Geographic, Washington, DC

Masters T (2014) Observational estimate of climate sensitivity from changes in the rate of ocean heat uptake and comparison to CMIP5 models. Clim Dyn 42(7):2173–2181. doi:10.1007/ s00382-013-1770-4

Masui T, Matsumoto K, Hijioka Y, Kinoshita T, Nozawa T, Ishiwatari S, Kato E, Shukla PR, Yamagata Y, Kainuma M (2011) An emission pathway for stabilization at 6 W m^{-2} radiative forcing. Clim Chang 109(1–2):59–76. doi:10.1007/s10584-011-0150-5

Medhaug I, Furevik T (2011) North Atlantic 20th century multidecadal variability in coupled climate models: sea surface temperature and ocean overturning circulation. Ocean Sci 7(3):389–404. doi:10.5194/os-7-389-2011

Meehl GA, Arblaster JM, Fasullo JT, Hu A, Trenberth KE (2011) Model-based evidence of deep-ocean heat uptake during surface-temperature hiatus periods. Nat Clim Chang 1(7):360–364. doi:10.1038/nclimate1229

Meinshausen M, Smith SJ, Calvin K, Daniel JS, Kainuma MLT, Lamarque JF, Matsumoto K, Montzka SA, Raper SCB, Riahi K, Thomson A, Velders GJM, Vuuren DPP (2011) The RCP greenhouse gas concentrations and their extensions from 1765 to 2300. Clim Chang 109(1–2):213–241. doi:10.1007/s10584-011-0156-z

Montzka SA, Dlugokencky EJ, Butler JH (2011) Non-CO_2 greenhouse gases and climate change. Nature 476(7358):43–50

Morgan MG, Adams PJ, Keith DW (2006) Elicitation of expert judgments of aerosol forcing. Clim Chang 75(1–2):195–214. doi:10.1007/s10584-005-9025-y

Myhre G, Highwood EJ, Shine KP, Stordal F (1998) New estimates of radiative forcing due to well mixed greenhouse gases. Geophys Res Lett 25:2715–2718. doi:10.1029/98GL01908

Norris JR, Allen RJ, Evan AT, Zelinka MD, O'Dell CW, Klein SA (2016) Evidence for climate change in the satellite cloud record. Nature 536:72–75. doi:10.1038/nature18273

Otto A, Otto FEL, Boucher O, Church J, Hegerl G, Forster PM, Gillett NP, Gregory J, Johnson GC, Knutti R, Lewis N, Lohmann U, Marotzke J, Myhre G, Shindell D, Stevens B, Allen MR

(2013) Energy budget constraints on climate response. Nat Geosci 6(6):415–416. doi:10.1038/ngeo1836

Peng J, Li Z, Zhang H, Liu J, Cribb M (2016) Systematic changes in cloud radiative forcing with aerosol loading for deep clouds in the tropics. J Atmos Sci 73(1):231–249. doi:10.1175/JAS-D-15-0080.1

Randall DA (2012) Atmosphere, clouds, and climate. Princeton primers in climate. Princeton University Press, Princeton

Raper SCB, Gregory JM, Stouffer RJ (2002) The role of climate sensitivity and ocean heat uptake on AOGCM transient temperature response. J Clim 15:124–130. doi:10.1175/1520-0442(2002)015<0124:TROCSA>2.0.CO;2

Riahi K, Rao S, Krey V, Cho C, Chirkov V, Fischer G, Kindermann G, Nakicenovic N, Rafaj P (2011) RCP 8.5—a scenario of comparatively high greenhouse gas emissions. Clim Chang 109(1–2):33–57. doi:10.1007/s10584-011-0149-y

Rieger LA, Bourassa AE, Degenstein DA (2015) Merging the OSIRIS and SAGE II stratospheric aerosol records. J Geophys Res Atmos 120(17):8890–8904. doi:10.1002/2015JD023133

Riser SC, Freeland HJ, Roemmich D, Wijffels S, Troisi A, Belbeoch M, Gilbert D, Xu J, Pouliquen S, Thresher A, Le Traon P-Y, Maze G, Klein B, Ravichandran M, Grant F, Poulain P-M, Suga T, Lim B, Sterl A, Sutton P, Mork K-A, Velez-Belchi PJ, Ansorge I, King B, Turton J, Baringer M, Jayne SR (2016) Fifteen years of ocean observations with the global Argo array. Nat Clim Change 6(2):145–153. doi:10.1038/nclimate2872

Rogelj J, den Elzen M, Höhne N, Fransen T, Fekete H, Winkler H, Schaeffer R, Sha F, Riahi K, Meinshausen M (2016) Paris Agreement climate proposals need a boost to keep warming well below 2 °C. Nature 534(7609):631–639. doi:10.1038/nature18307, http://www.nature.com/nature/journal/v534/n7609/abs/nature18307.html#supplementary-information

Rypdal K (2015) Attribution in the presence of a long-memory climate response. Earth Syst Dyn 6(2):719–730. doi:10.5194/esd-6-719-2015

Saji HH, Goswami BN, Vinayachandran PH, Yamagata T (1999) A dipole mode in the tropical Indian Ocean. Nature 401:360–363. doi:10.1038/43854

Santer BD, Painter JF, Bonfils C, Mears CA, Solomon S, Wigley TM, Gleckler PJ, Schmidt GA, Doutriaux C, Gillett NP, Taylor KE, Thorne PW, Wentz FJ (2013a) Human and natural influences on the changing thermal structure of the atmosphere. Proc Natl Acad Sci U S A 110(43):17235–17240. doi:10.1073/pnas.1305332110

Santer BD, Painter JF, Mears CA, Doutriaux C, Caldwell P, Arblaster JM, Cameron-Smith PJ, Gillett NP, Gleckler PJ, Lanzante J, Perlwitz J, Solomon S, Stott PA, Taylor KE, Terray L, Thorne PW, Wehner MF, Wentz FJ, Wigley TM, Wilcox LJ, Zou CZ (2013b) Identifying human influences on atmospheric temperature. Proc Natl Acad Sci U S A 110(1):26–33. doi:10.1073/pnas.1210514109

Santer BD, Bonfils C, Painter JF, Zelinka MD, Mears CA, Solomon S, Schmidt GA, Fyfe JC, Cole JNS, Nazarenko L, Taylor KE, Wentz FJ (2014) Volcanic contribution to decadal changes in tropospheric temperature. Nat Geosci 7(3):185–189. doi:10.1038/ngeo2098

Saravanan R, McWilliams JC (1998) Advective ocean–atmosphere interaction: an analytical stochastic model with implications for decadal variability. J Clim 11:165–188

Sato M, Hansen JE, McCormick MP, Pollack JB (1993) Stratospheric aerosol optical depths, 1850–1990. J Geophys Res 98:22987–22994. doi:10.1029/93JD02553

Schlesinger ME, Ramankutty N (1994) An oscillation in the global climate system of period 65–70 years. Nature 367(6465):723–726. doi:10.1038/367723a0

Schmidt GA, Shindell DT, Tsigaridis K (2014) Reconciling warming trends. Nat Geosci 7(3):158–160. doi:10.1038/ngeo2105

Schwartz SE (2012) Determination of earth's transient and equilibrium climate sensitivities from observations over the twentieth century: strong dependence on assumed forcing. Surv Geophys 33(3-4):745–777. doi:10.1007/s10712-012-9180-4

Shindell DT, Lamarque JF, Schulz M, Flanner M, Jiao C, Chin M, Young PJ, Lee YH, Rotstayn L, Mahowald N, Milly G, Faluvegi G, Balkanski Y, Collins WJ, Conley AJ, Dalsoren S, Easter R,

Ghan S, Horowitz L, Liu X, Myhre G, Nagashima T, Naik V, Rumbold ST, Skeie R, Sudo K, Szopa S, Takemura T, Voulgarakis A, Yoon JH, Lo F (2013) Radiative forcing in the ACCMIP historical and future climate simulations. Atmos Chem Phys 13(6):2939–2974. doi:10.5194/acp-13-2939-2013

Silver N (2012) The signal and the noise: why so many predictions fail—but some don't. Penguin Press, New York

Smith SJ, Bond TC (2014) Two hundred fifty years of aerosols and climate: the end of the age of aerosols. Atmos Chem Phys 14(2):537–549. doi:10.5194/acp-14-537-2014

Smith SJ, van Aardenne J, Klimont Z, Andres RJ, Volke A, Delgado Arias S (2011) Anthropogenic sulfur dioxide emissions: 1850–2005. Atmos Chem Phys 11(3):1101–1116. doi:10.5194/acp-11-1101-2011

Solomon S, Daniel JS, Neely RR III, Vernier JP, Dutton EG, Thomason LW (2011) The persistently variable "background" stratospheric aerosol layer and global climate change. Science 333(6044):866–870. doi:10.1126/science.1206027

Solomon S, Ivy DJ, Kinnison D, Mills MJ, Neely RR, Schmidt A (2016) Emergence of healing in the Antarctic ozone layer. Science 353(6296):269–274. doi:10.1126/science.aae0061

Srokosz MA, Bryden HL (2015) Observing the Atlantic Meridional Overturning Circulation yields a decade of inevitable surprises. Science 348(6241):1255575. doi:10.1126/science.1255575

Stern DI (2006a) An atmosphere-ocean time series model of global climate change. Comput Stat Data Anal 51(2):1330–1346. doi:10.1016/j.csda.2005.09.016

Stern DI (2006b) Reversal of the trend in global anthropogenic sulfur emissions. Glob Environ Chang 16(2):207–220. doi:10.1016/j.gloenvcha.2006.01.001

Stern DI, Kaufmann RK (2014) Anthropogenic and natural causes of climate change. Clim Chang 122(1):257–269. doi:10.1007/s10584-013-1007-x

Storelvmo T, Lohmann U, Bennartz R (2009) What governs the spread in shortwave forcings in the transient IPCC AR4 models? Geophys Res Lett 36(1):L01806. doi:10.1029/2008GL036069

Stott P, Good P, Jones G, Gillett N, Hawkins E (2013) The upper end of climate model temperature projections is inconsistent with past warming. Environ Res Lett 8(1):014024

Stouffer RJ, Yin J, Gregory JM, Dixon KW, Spelman MJ, Hurlin W, Weaver AJ, Eby M, Flato GM, Hasumi H, Hu A, Jungclaus JH, Kamenkovich IV, Levermann A, Montoya M, Murakami S, Nawrath S, Oka A, Peltier WR, Robitaille DY, Sokolov A, Vettoretti G, Webber SL (2006) Investigating the causes of the response of the thermohaline circulation to past and future climate changes. J Clim 19:1365–1387. doi:10.1175/JCLI3689.1

Taylor JR (1982) An introduction to error analysis: the study of uncertainties in physical measurements. A series of books in physics. University Science Books, Mill Valley, CA

Taylor KE, Stouffer RJ, Meehl GA (2012) An overview of CMIP5 and the experiment design. Bull Am Meteorol Soc 93(4):485–498. doi:10.1175/bams-d-11-00094.1

Thompson DWJ, Wallace JM, Jones PD, Kennedy JJ (2009) Identifying signatures of natural climate variability in time series of global-mean surface temperature: methodology and insights. J Clim 22(22):6120–6141. doi:10.1175/2009jcli3089.1

Thomson AM, Calvin KV, Smith SJ, Kyle GP, Volke A, Patel P, Delgado-Arias S, Bond-Lamberty B, Wise MA, Clarke LE, Edmonds JA (2011) RCP4.5: a pathway for stabilization of radiative forcing by 2100. Clim Chang 109(1–2):77–94. doi:10.1007/s10584-011-0151-4

Trenberth KE, Fasullo JT (2013) An apparent hiatus in global warming? Earth's Future 1(1):19–32. doi:10.1002/2013EF000165

van Vuuren DP, Stehfest E, Elzen MGJ, Kram T, Vliet J, Deetman S, Isaac M, Klein Goldewijk K, Hof A, Mendoza Beltran A, Oostenrijk R, Ruijven B (2011a) RCP2.6: exploring the possibility to keep global mean temperature increase below 2 °C. Clim Chang 109(1–2):95–116. doi:10.1007/s10584-011-0152-3

van Vuuren DP, Edmonds J, Kainuma M, Riahi K, Thomson A, Hibbard K, Hurtt GC, Kram T, Krey V, Lamarque J-F, Masui T, Meinshausen M, Nakicenovic N, Smith SJ, Rose SK (2011b)

The representative concentration pathways: an overview. Clim Chang 109(1–2):5–31. doi:10.1007/s10584-011-0148-z

Vial J, Dufresne J-L, Bony S (2013) On the interpretation of inter-model spread in CMIP5 climate sensitivity estimates. Clim Dyn 41(11–12):3339–3362. doi:10.1007/s00382-013-1725-9

Vincze M, Jánosi IM (2011) Is the Atlantic Multidecadal Oscillation (AMO) a statistical phantom? Nonlinear Process Geophys 18(4):469–475. doi:10.5194/npg-18-469-2011

Wang YM, Lean JL, Sheeley NR Jr (2005) Modeling the Sun's magnetic field and irradiance since 1713. Astrophys J 625:522–538. doi:10.1086/429689

Weaver C, Herman J, Labow G, Larko D, Huang L-K (2015) Shortwave TOA cloud radiative forcing derived from a long-term (1980–present) record of satellite UV reflectivity and CERES measurements. J Clim 28(23):9473–9488. doi:10.1175/JCLI-D-14-00551.1

Webb MJ, Andrews T, Bodas-Salcedo A, Bony S, Bretherton CS, Chadwick R, Chepfer H, Douville H, Good P, Kay JE, Klein SA, Marchand R, Medeiros B, Siebesma AP, Skinner CB, Stevens B, Tselioudis G, Tsushima Y, Watanabe M (2016) The Cloud Feedback Model Intercomparison Project (CFMIP) contribution to CMIP6. Geosci Model Dev Discuss 2016:1–27. doi:10.5194/gmd-2016-70

Willis JK (2010) Can in situ floats and satellite altimeters detect long-term changes in Atlantic Ocean overturning? Geophys Res Lett 37(6):L06602. doi:10.1029/2010GL042372

Wu L, Liu Z (2003) Decadal variability in the North Pacific: the Eastern North Pacific mode. J Clim 16:3111–3131

Zelinka MD, Klein SA, Taylor KE, Andrews T, Webb MJ, Gregory JM, Forster PM (2013) Contributions of different cloud types to feedbacks and rapid adjustments in CMIP5. J Clim 26(14):5007–5027. doi:10.1175/jcli-d-12-00555.1

Zhang Y, Wallace JM, Battisti DS (1997) ENSO-like interdecadal variability: 1900–93. J Clim 10:1004–1020

Zhang R, Delworth TL, Held IM (2007) Can the Atlantic Ocean drive the observed multidecadal variability in Northern Hemisphere mean temperature? Geophys Res Lett 34(2):L02709. doi:10.1029/2006GL028683

Zhou J, Tung K-K (2013) Deducing multidecadal anthropogenic global warming trends using multiple regression analysis. J Atmos Sci 70(1):3–8. doi:10.1175/jas-d-12-0208.1

Zhou C, Zelinka MD, Dessler AE, Klein SA (2015) The relationship between interannual and long-term cloud feedbacks. Geophys Res Lett 42(23):10,463–410,469. doi:10.1002/2015GL066698

Chapter 3
Paris INDCs

**Walter R. Tribett, Ross J. Salawitch, Austin P. Hope,
Timothy P. Canty, and Brian F. Bennett**

Abstract This chapter begins with a description of the Paris Climate Agreement, which was formulated during the 21st meeting of the Conference of the Parties to the United Nations Framework Convention on Climate Change (UNFCCC) in late 2015. The goal of this agreement is to limit future emission of greenhouse gases (GHGs) such that global warming will not exceed 1.5 °C (target) or 2.0 °C (upper limit). Future emissions of GHGs are based on unilateral pledges submitted by UNFCCC member nations, called Intended Nationally Determined Contributions (INDCs). We compare the global emission of GHGs calculated from the INDCs to the emissions that had been used to formulate the various Representative Concentration Pathway (RCP) trajectories for future atmospheric abundance of GHGs. The RCP 4.5 scenario is particularly important, because our Empirical Model of Global Climate (EM-GC) indicates there is a reasonably good probability (~75 %) the Paris target will be achieved, and an excellent probability (>95 %) the upper limit for global warming will be attained, if the future atmospheric abundance of GHGs follows RCP 4.5. Our analysis of the Paris INDCs shows GHG emissions could remain below RCP 4.5 out to year 2060 if: (1) conditional as well as unconditional INDCs are followed; (2) reductions in GHG emissions needed to achieve the Paris INDC commitments, which generally stop at 2030, are propagated forward to 2060. Prior and future emissions of GHGs are graphically illustrated to provide context for the reductions needed to place global GHG emissions on the RCP 4.5 trajectory.

Keywords Paris Climate Agreement • Paris INDCs • Greenhouse gas emissions • CO_2-equivalent emissions • Unconditional INDC • Conditional INDC

3.1   Introduction

The Paris Climate Agreement has a structure distinctly different than its predecessor, the Kyoto Protocol. The Kyoto Protocol was approved at the third meeting of the Conference of the Parties (COP) of the United Nations Framework Convention on Climate Change (UNFCCC) held in Kyoto, Japan during December, 1997. The goal of Kyoto was to minimize the adverse effects of climate change due to rising

© The Author(s) 2017
R.J. Salawitch et al., *Paris Climate Agreement: Beacon of Hope*,
Springer Climate, DOI 10.1007/978-3-319-46939-3_3

Table 3.1 Annex I nations of
the Kyoto Protocol

Australia	Greece	Norway
Austria	Hungary	Poland
Belarus	Iceland	Portugal
Belgium	Ireland	Romania
Bulgaria	Italy	Russia
Canada	Japan	Slovakia
Croatia	Latvia	Slovenia
Cyprus	Liechtenstein	Spain
Czech Republic	Lithuania	Sweden
Denmark	Luxembourg	Switzerland
Estonia	Malta	Turkey
Finland	Monaco	Ukraine
France	Netherlands	United Kingdom
Germany	New Zealand	United States

levels of greenhouse gases (GHGs).[1] The governing document focused on reducing emissions of carbon dioxide (CO_2), methane (CH_4), nitrous oxide (N_2O), hydrofluorocarbons (HFCs), perfluorocarbons (PFCs), sulfur hexafluoride (SF_6) (Article 3), known as the Kyoto basket of GHGs. The world was split into two categories: Annex I nations (Table 3.1) and the rest of the world, which we refer to as the Non-Annex I nations. The Annex I nations consist of what most would have considered to be a reasonably good representation of the developed world in 1997.

According to the terms of the Protocol, Annex I nations had varying emission reduction targets for the Kyoto basket of GHGs, relative to emissions in year 1990 from that particular country. Total emissions of GHGs were combined into a single emission metric, termed CO_2-equivalent (CO_2-eq) emission, attained by multiplying the annual emissions of each compound by the global warming potential of that compound.[2] Each Annex I signatory nation negotiated an emission reduction target, except that the 15 European nations agreed to follow a single, combined target referred to as EU15. The target for the US was a 7 % reduction in CO_2-eq emissions and the EU15 target was an 8 % reduction, both to be achieved by 2005 relative to 1990. The largest reduction was 8 % (shared by several other nations in addition to EU15). Some signatories were allowed to increase emissions at a prescribed limit to growth, such as

---

[1] See http://unfccc.int/essential_background/kyoto_protocol/items/1678.php for the actual document; versions in many other languages at http://unfccc.int/kyoto_protocol/items/2830.php

[2] Typically, emissions are quantified as mass of each compound released over a year, and GWPs are based on the use of a 100-year time horizon for the governing equation. By definition, the GWP for CO_2, regardless of the source, is unity (i.e., equals 1). Numerous complications arise from the CO_2-equivalence convention, most notably the fact that best-estimates of GWPs change over time (Table 1.2), and often papers and reports do not document which GWP was actually used. Throughout this book, we use GWPs for CH_4 and N_2O of 28 and 265, respectively, unless otherwise stated. Another complication is that the effect of inadvertent release of CH_4 on global warming over the decadal time scale is not properly represented by the use of GWP on a 100-year time horizon, as discussed in Sects. 1.2.2 and 4.4.2, as well as by Pierrehumbert (2014).

Australia which agreed to have CO_2-eq emissions in 2005 be no more than 8 % larger than had occurred in 1990. The highest increase allowed was 10 %, for Iceland.

A few more pertinent details of the Kyoto Protocol follow. The Protocol allowed nations to use reductions in CO_2 attributed to land use change (LUC) to meet their commitment,[3] provided the decline in emission occurred due to direct human induced LUC since 1990. Kyoto included three mechanisms to assist nations in meeting their targets: Joint Implementation,[4] Clean Development,[5] and Emissions Trading.[6] If a country or the EU15 group failed to achieve their target during the first commitment period, which ended in 2012, two consequences ensued: a 30 % penalty of additional emission reductions for the second commitment period, and suspension of the ability to sell emissions trading credits. Details of the Kyoto Protocol were continually refined at subsequent meetings of the UNFCCC COP, held annually towards the end of the calendar year.[7]

There has been so much written about the Kyoto Protocol that references hardly seem necessary. At the time of writing, the Amazon website returns 5011 results for a search on "Kyoto protocol" in Books. We do, however, suggest *The Collapse of the Kyoto Protocol and the Struggle to Slow Global Warming* (Victor 2001) as a concise and accessible account of this agreement and its subsequent amendments, including thoughtful exposition about positive aspects of the Protocol as well as suggestions for what could have been done better.

The Kyoto Protocol did not place restrictions on GHG emissions from developing countries (i.e., all countries not listed in Table 3.1). A sub-group of Annex I nations, termed Annex II and consisting of Australia, Austria, Belgium, Canada, Denmark, Finland, France, Germany, Greece, Iceland, Ireland, Italy, Japan, Luxembourg, Netherlands, New Zealand, Norway, Portugal, Spain, Sweden, Switzerland, Turkey, the United Kingdom, and the United States, were tasked with providing financial support for the development of technology to reduce GHG emissions in developing countries.

At some point in time, the Kyoto Protocol had been signed and ratified by all nations except Afghanistan, Southern Sudan, Taiwan, and the United States.[8] Canada withdrew from Kyoto in 2011, due to perceived pressure on the extraction of bitumen from Canadian tar sands. The US Congress failed to ratify the Protocol,

---

[3] The official language for LUC in the Protocol calls this land use, land-use change and forestry and uses the abbreviation of LULUCF. Here and throughout, we use the more simple abbreviation LUC, with recognition of the importance of forestry.

[4] Joint implementation allowed Annex I countries to implement projects that reduce emissions or increase natural GHG sinks in other Annex I countries; such projects could be counted towards the emission reductions of the investing country.

[5] Clean Development allows Annex I countries to implement projects that reduce emissions or increase natural GHG sinks in non-Annex I countries; such projects can be counted towards the emission reductions of the investing country.

[6] Annex I countries could purchase emission units from other Annex I countries that found it easier to reduce their own emissions.

[7] A UNFCCC COP schedule is at http://unfccc.int/meetings/items/6240.php

[8] Observer nations Andorra and Vatican City are sometimes listed as non-participants, but their emissions are too small to matter, plus Vatican City answers to a higher authority.

which required Congressional Approval because it was viewed as a treaty by the US Government. In fact, on 25 July 1997 the Senate of the 105th Congress approved, by a vote of 95 to 0, a resolution[9] that declared:

> the United States should not be a signatory to any protocol to, or other agreement regarding, the United Nations Framework Convention on Climate Change of 1992, at negotiations in Kyoto in December 1997 or thereafter which would: (1) mandate new commitments to limit or reduce greenhouse gas emissions for the Annex 1 Parties, unless the protocol or other agreement also mandates new specific scheduled commitments to limit or reduce greenhouse gas emissions for Developing Country Parties within the same compliance period; or (2) result in serious harm to the US economy.

This resolution was passed *six months prior* to the Kyoto meeting. Since the Protocol did not include "specific scheduled commitments" to limit GHG reductions from developing countries, approval by the US Congress was always going to be an uphill battle (Victor 2001; Falkner et al. 2010).

On 12 November 2014, nearly 20 years after the Kyoto meeting, President Obama of the US and President Xi of China announced a set of crucially important, bilateral GHG reduction targets.[10] According to their announcement, by 2025 the US would reduce its total GHG emissions to be 26–28 % below the total emission that had occurred in 2005. China agreed to have their CO_2 emissions peak by 2030 and to make best effort to peak early. China also stated it would increase its share of the use of non-fossil fuels in its primary energy consumption to about 20 % by 2030. There were a variety of other actions, such as joint efforts to phase down the global use of HFCs, a class of GHGs introduced by the ban on chlorofluorocarbons to comply with the Montreal Protocol (Velders et al. 2007; see also Sect. 1.2.3.5), promote energy efficiency in buildings, and support research into carbon capture and sequestration (CCS) technologies (Sect. 4.2).

The structure of the Paris Climate Agreement is quite different than that of the Kyoto Protocol. First and foremost, the Paris Agreement has specific goals for limiting future global warming relative to the pre-industrial baseline. The Agreement[11] seeks to reduce cumulative emission of GHGs such that the increase in global mean surface temperature (GMST) is "well below 2 °C" and to "pursue efforts to limit the temperature increase to 1.5 °C above pre-industrial". Throughout this book, we have interpreted these two numbers as being the "Paris target of 1.5 °C warming" and the "Paris upper limit of 2.0 °C warming".

The second aspect of the Paris Agreement that differs from the Kyoto Protocol is that individual nations were encouraged to submit, prior to the COP 21 meeting in Paris, their unilateral Intended Nationally Determined Contribution (INDC) for the reduction of GHG emissions. There are two types of INDCs: unconditional (firm commitments) and conditional (commitments contingent on financial assistance and/or technology transfer). The INDCs from most participating nations in the

---

[9] https://www.congress.gov/bill/105th-congress/senate-resolution/98

[10] https://www.whitehouse.gov/the-press-office/2014/11/11/us-china-joint-announcement-climate-change

[11] English language version at http://unfccc.int/files/essential_background/convention/application/pdf/english_paris_agreement.pdf

developing world are conditional. The Green Climate Fund (Sect. 4.3), established during COP 15, is recognized as one of several means to facilitate the flow of resources needed to implement the conditional INDCs. The Paris INDCs consider the original Kyoto basket of GHGs (CO_2, CH_4, N_2O, HFCs, PFCs, and SF_6) plus NF_3, which was added at the COP 17 meeting held in Durban, South Africa during 2011.[12] Below, we refer to this group of seven as the UNFCCC basket of GHGs.

The Obama-Xi announcement was instrumental in the framing of the Paris Climate Agreement. The INDCs submitted by the US and China, both unconditional, build closely on the language of this bilateral plan. These nations emit more GHGs than any other: China bypassed the US to become the world's largest emitter of CO_2 during 2006. The importance of these two nations arriving at mutually agreeable language to combat global warming, prior to the Paris meeting, cannot be understated. To date, INDCs from 190 out of the 196 nations in the world have been submitted to UNFCCC. For the first time in history, there is consensus among the world's nations that a collective effort is needed to combat global warming.

Much will be written comparing and contrasting the Paris Climate Agreement and the Kyoto Protocol. The Paris Climate Agreement has a top-down, quantitative goal of limiting global warming from rising either 1.5 °C (target) or 2.0 °C (upper limit) above pre-industrial. The method of achieving the necessary reduction in GHG emissions is a bottom-up approach, conducted via unilateral INDCs. The Obama administration maintains the agreement is not a treaty and, as such, does not require Congressional approval. The Obama administration has proposed to fulfill the US commitment via the Clean Power Plan, an Environmental Protection Agency proposal to limit the emission of CO_2 from power plants within each of the 50 states (Sect. 4.4.2).

An overview of the historical emission of GHGs is provided in Sect. 3.2. Agreements such as Paris do not occur in a vacuum: i.e., an enormous amount of effort takes place prior to each COP meeting. Past emissions, economic resources, technology, and each nation's perspective on environmental responsibility play large roles in the framing of the guiding document as well as the content of individual INDCs. Past emissions of GHGs are illustrated in Sect. 3.2, both globally and nationally, because these data are readily available and provide an interesting backdrop to the Paris Climate Agreement.

Global emissions of GHGs implied by the Paris INDCs are quantified in Sect. 3.3. Projected emissions of GHGs inferred from the Paris INDCs are compared to the emissions that were used to drive the RCP 8.5 (Riahi et al. 2011), RCP 4.5 (Thomson et al. 2011), and RCP 2.6 (van Vuuren et al. 2011) scenarios, which are central to IPCC (2013). The RCP 4.5 scenario is a particularly important benchmark. Calculations shown in Chap. 2, conducted using our Empirical Model of Global Climate (EM-GC) (Canty et al. 2013), indicate there is a reasonably high probability (~75 %) that the Paris target of 1.5 °C warming will be achieved, and an excellent probability (>95 %) that global warming will remain below 2.0 °C, if the

---

[12] http://unfccc.int/press/news_room/newsletter/in_focus/items/6672.php. The decision to add NF_3 to the Kyoto basket was made at Durban, South Africa in 2011, this GHG was formally added via an amendment to the protocol approved in Doha, Qatar in 2012. More information about NF_3 is given in Sect. 1.2.3.5.

atmospheric abundance of GHGs follows RCP 4.5. Conversely, there is little to no chance these warming limits will be achieved if emissions follow RCP 8.5.

Our evaluation of GHG emissions comes with an important condition as well as a crucial caveat. The condition is that, to properly evaluate the Paris Agreement, emissions of GHGs must be examined at least out to year 2060. Most of the INDCs extend only to year 2030. As shown in Sects. 3.3 and 4.2, the 2030–2060 time period is crucial. Assuming populations continue to grow and standards of living continue to rise as projected, then the production of a large amount of total global energy by methods that release little or no atmospheric GHGs by 2060 will be vital for the achievement of the Paris Agreement. While it is tempting to extend the comparison of GHG emission projections out to 2100, it is not realistic to consider policy measures out to end of century. However, power plants commissioned during the next decade will almost certainly be designed to be operational in 2030. As shown in Sect. 4.2, for the world to achieve the reduction in GHG emissions needed to lie along the RCP 4.5 trajectory in 2060, we must meet about half of the projected global demand for energy without releasing GHGs to the atmosphere. For this to happen, it is incumbent that planning begin now.

The crucial caveat of our projections is that use of RCP 4.5 as the benchmark for evaluating the Paris Agreement depends on the veracity of the calculations conducted using our EM-GC framework. The coupled atmospheric, oceanic general circulation models (GCMs) used extensively by IPCC (2013) indicate that the RCP 2.6 scenario (van Vuuren et al. 2011), which imposes much tighter constraints on GHG emissions than RCP 4.5, is the appropriate benchmark for Paris (Rogelj et al. 2016). In Chap. 2, values of the Attributable Anthropogenic Warming Rate (AAWR) inferred from the climate record were compared to AAWR from GCMs. We concluded that GCMs tend to warm too quickly, by a rate that exceeds the observed warming rate by nearly a factor of two. Our conclusion that GCMs warm too quickly is consistent with the findings of Chap. 11 of IPCC (2013), particularly their expert judgement of projected warming over the next two decades that plays a prominent role in our Chap. 2.

The global warming target (1.5 °C) and upper limit (2.0 °C) of the Paris Climate Agreement will undoubtedly spur many other evaluations of GCMs, as well as other empirical forecasts of global warming. If the consensus of this research demonstrates that RCP 2.6 is indeed a more appropriate benchmark for achieving the goal of Paris than RCP 4.5, then GHG emissions will need to be reduced much faster than in the present INDC commitments to have any hope of achieving either the target or upper limit of the Paris Climate Agreement (Rogelj et al. 2016; see also Sect. 4.2).

3.2 Prior Emissions

Here, an overview of the historical emission of GHGs is provided. Numerous papers, reports, and blogs focus solely on emissions of CO_2 (Pacala and Socolow 2004; Canadell et al. 2007; Raupach et al. 2007; Friedlingstein et al. 2014), in most cases due only to the combustion of fossil fuels. However, the Paris Climate Agreement covers the UNFCCC basket of GHGs, and CO_2 emission from land use change, in addition to fossil fuels. As shown below, the average global, per-capita emission of

CO_2, CH_4, and N_2O summed among all human sources is about 7.5 metric tons of CO_2-eq per person per year. Conversely, the global, per-capita emission of CO_2 due to the combustion of fossil fuels is about 5 metric tons of CO_2 per person per year. Adding CH_4, N_2O, and CO_2 from LUC to the mix requires even steeper cuts in total GHG emissions to achieve the goals of the Paris Climate Agreement than would be needed if the focus were solely on release of CO_2 from combustion of fossil fuels.

Prior emission of GHGs by individual nations played an important role in the framing of the Paris Climate Agreement. Many of the INDCs use language that makes specific reference to prior emissions. We therefore show maps of national emissions of GHGs, presented in terms of CO_2 from the combustion of fossil fuels as well as human emission of CO_2, CH_4, and N_2O from all sources.

In the material that follows, our focus is solely on anthropogenic emission of CO_2, CH_4, and N_2O. This is not to diminish the importance of other GHGs, as well as other human drivers of climate change such as rising tropospheric O_3 and industrial release of CFCs and other ozone depleting substances (Fig. 1.4). We neglect tropospheric O_3 here because the precursors of tropospheric O_3 are regulated by Air Quality policy makers rather than the climate community. We neglect ozone depleting substances because these compounds are regulated, quite effectively, by the Montreal Protocol (Sect. 1.2.3.4). And, we do not discuss other fluorine-bearing GHGs such as HFCs, PFCs, SF_6, and NF_3 because, to date, their contribution to the RF of climate has been small (Fig. 1.4). Projections of the future radiative impacts of HFCs, PFCs, SF_6, and NF_3, due to market forces independent of the Paris Climate Agreement, are discussed in Sect. 1.2.3.5. The climate impact of HFCs could be considerable in the future, particularly if compounds with extremely high GWPs are left unregulated (Velders et al. 2009). As discussed in Chap. 1, future regulation of HFCs has recently been approved by the Parties of the Montreal Protocol. Given this effort, plus the very minor role attributed to SF_6, PFCs, and HFCs out to 2060 in the RCP projections, it seemed prudent to restrict our focus to the big three: CO_2, CH_4, and N_2O.

3.2.1   Global

Figure 3.1a illustrates global, annual emission of atmospheric CO_2 from the combustion of fossil fuels, over the prior two centuries. As noted in Sect. 1.2.3.2, about half of the CO_2 released to the atmosphere by human activity remains airborne, while the rest is removed by either the world's oceans or terrestrial biosphere (mainly trees). This figure shows the total global, annual emission of atmospheric CO_2 from the combustion of coal, natural gas, liquid fuels, cement manufacture, and gas flaring (CO_2^{FF}), obtained from the US Carbon Dioxide Information and Analysis Center (CDIAC)(Boden et al. 2013; Le Quéré et al. 2015). Data are shown in units of Gt CO_2 per year.[13] Global population is also shown.

---

[13] 1 Gt of $CO_2 = 10^9$ metric tons of CO_2. Emissions of CO_2 are expressed either as Gt C or Gt CO_2. Emissions given in Gt C can be converted to Gt CO_2 by multiplying the value by 3.664 (Table 1 of Le Quéré et al. (2015)).

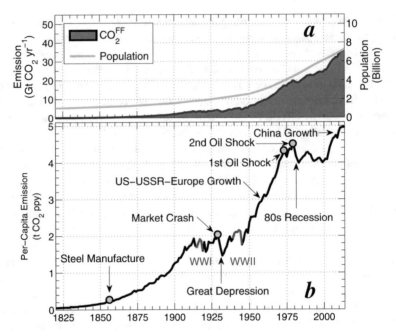

Fig. 3.1 Total global emission of atmospheric CO_2. (**a**) Emission of CO_2 from combustion of fossil fuels, flaring, and cement manufacture (CO_2^{FF}, *grey shaded*) as well as global population (*green*), from 1820 to 2014; (**b**) per-capita emission of global atmospheric CO_2 (pC^{GL}) expressed in metric tons of CO_2 per person, per year (t CO_2 ppy). World events associated with changes in pC^{GL} are noted. See Methods for further information

Figure 3.1b shows global, per-capita emission of CO_2 from the combustion of fossil fuel,[14] which we abbreviate as pC^{GL}. Values of pC^{GL} are presented in units of metric tons of CO_2 per person per year, abbreviated as t CO_2 ppy. There was a steady rise in pC^{GL} from 1856, which marks the beginning of the mass production of steel (Adams and Dirlam 1966), until the start of World War I. A hiatus in pC^{GL} then occurred until the end of World War II, followed by a rapid rise until 1973. Most of this growth drove the economic development of the US, Europe, and the former USSR. Many attribute the abrupt leveling off of pC^{GL} in 1973 to the rapid rise in the price of oil that followed the 6-day Yom Kippur war between Egypt and Israel (first Oil Shock) (Hamilton 2003). This second hiatus in pC^{GL} lasted until 2000. During this 27 year period, there was a series of world events, such as a second rapid rise in the price of oil driven by the Iranian revolution (second Oil Shock) and the 1980s economic recession, all of which contributed to significant increases in carbon efficiency within the developed world. Since 2002, the economic development of China has led to a third period marked by a rise in pC^{GL} (Le Quéré et al. 2015). It is remarkable how many world events are apparent in the record of per-capita con-

---

[14] Per-capita equals global emissions divided by global population; the work capita has Latin roots, meaning head.

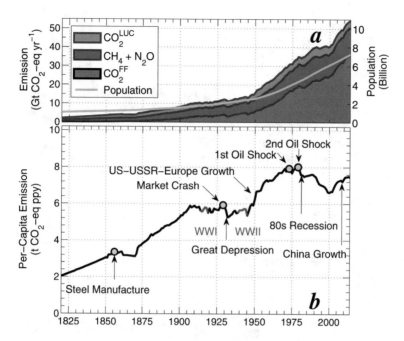

Fig. 3.2 Total global emission of atmospheric CO_2, CH_4, and N_2O. (**a**) Emission of CO_2 from combustion of fossil fuels (CO_2^{FF}; same as Fig. 3.1), anthropogenic emission of CH_4 plus N_2O expressed as CO_2-equivalent (CO_2-eq) (*blue*), and emission of CO_2 from land use change (CO_2^{LUC}, *red*); global population (*green*) is also shown; (**b**) per-capita emission of CO_2^{FF} + CO_2^{LUC} + CH_4 + N_2O, termed pC^{EQ-GL}, expressed in metric tons of CO_2-eq per person, per year (t CO_2-eq ppy). World events associated with changes in pC^{EQ-GL} are noted. See Methods for further information

sumption of fossil fuel, which has had two distinct growth spurts (1860–1910; 1950–1973) and appears to be entering a third period of growth.

The Paris INDCs focus on reducing the emission of the UNFCCC basket of GHGs, expressed in terms of CO_2-eq (Sect. 3.1). Release of CO_2 by the combustion of fossil fuel is the most important contributor to this total GHG emission burden. Total anthropogenic emission of CH_4, which is released to the atmosphere by many aspects of our industrialized world (Sect. 1.2.3.3), is the second largest contributor. The release of CO_2 by land use change (CO_2^{LUC}) and the emission of N_2O (Sect. 1.2.3.4) make additional contributions, nearly equal in magnitude, that must be considered when examining the Paris INDCs.

Figure 3.2a shows a time series of CO_2-eq emission of GHGs. The four most important terms are included: CO_2^{FF}, CO_2^{LUC}, CH_4, and N_2O. Global population is also shown in Fig. 3.2a. The per-capita emission of GHGs in the Paris INDC relevant metric, CO_2-eq, is shown in Fig. 3.2b. The quotient of CO_2-eq emissions divided by global population, which reflects the globally averaged contribution to global warming by the world's population, is termed pC^{EQ-GL}.

Figure 3.3 shows the breakdown of anthropogenic release of CH_4 and N_2O, in CO_2-eq units. Figure 3.3a shows emission estimates for CH_4 and N_2O from the same source, RCP (Meinshausen et al. 2011), which has been the resource used for global

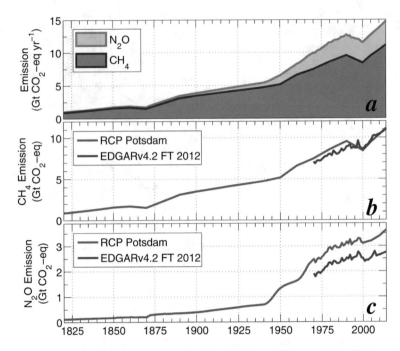

Fig. 3.3 Total global emissions of atmospheric CH$_4$ and N$_2$O. (**a**) Emission of CH$_4$ (*blue*) and N$_2$O (*green*) expressed as CO$_2$-equivalent (CO$_2$-eq) from the RCP Potsdam database (Meinshausen et al. 2011); (**b**) comparison of global emissions of CH$_4$ from the RCP Potsdam database and from the Emissions Database for Global Atmospheric Research (EDGAR) version 4.2 FT2012 database (Rogelj et al. 2014); (**c**) same as panel (**b**), but for N$_2$O. See Methods for further information

emission of GHGs throughout this book. In this case, the two estimates represent an attempt to harmonize the emissions used to drive the RCP scenarios with atmospheric observations of CH$_4$ and N$_2$O.[15] Emissions of CH$_4$, expressed as CO$_2$-eq using a 100-year time horizon, constitute about 80 % of the sum.

Figures 3.3b, c compare the RCP emissions of CH$_4$ and N$_2$O, respectively, to values from the Emissions Database for Global Atmospheric Research (EDGAR)[16] database (Rogelj et al. 2014). The two estimates for CH$_4$ are in very good agreement. However, they both use similar (or perhaps the same) measurements of the atmospheric abundance of CH$_4$ versus time (Fig. 2.1) to guide the respective time series. Both emission time series show that human release of CH$_4$ appears to have stalled in the 1990s, before pickup up in the most recent decade. The precise reason for this behavior is the subject of considerable uncertainty, perhaps best summarized by Kirschke et al. (2013), who in their abstract state:

> Although uncertainties in emission trends do not allow definitive conclusions to be drawn, we show that the observed stabilization of methane levels between 1999 and 2006 can potentially be explained by decreasing-to-stable fossil fuel emissions, combined with stable-to-

---

[15] See http://www.pik-potsdam.de/~mmalte/rcps for further information.

[16] Here and throughout, we use version 4.2 FT 2012 emissions from EDGAR.

increasing microbial emissions. We show that a rise in natural wetland emissions and fossil fuel emissions probably accounts for the renewed increase in global methane levels after 2006, although the relative contribution of these two sources remains uncertain.

Figure 3.3c compares the RCP (Meinshausen et al. 2011) and EDGAR (Rogelj et al. 2014) estimates of the global emission of N_2O. Clearly there are common roots to these two estimates, based on the synchronization of the fluctuations. However, the RCP estimate exceeds the EDGAR by about 1 Gt CO_2-eq, for reasons that are unclear.

The emissions of CH_4 and N_2O from EDGAR and RCP have been compared in Fig. 3.3 because of their complementary importance to this book. The emissions from RCP, which are provided globally, extend back to 1765 (Meinshausen et al. 2011). This allows the historical evolution of the most important subset of the UNFCCC basket of GHGs (i.e., CO_2, CH_4, and N_2O) to be examined over the past two centuries (Fig. 3.2). Conversely, the emissions from EDGAR extend back to 1970. However, EDGAR documents national emissions of CH_4 and N_2O for each year, from 1970 to present. This is vitally important information for assessing national burdens towards global warming, as well as the evaluating the Paris INDCs.

We now turn our attention to comparing and contrasting the time series of per-capita emission of CO_2 from the combustion of fossil fuels (pC^{GL}) (Fig. 3.2a) with per-capita emission of all human sources of CO_2, CH_4, and N_2O (pC^{EQ-GL}) (Fig. 3.2b). Most of the world events are still evident in pC^{EQ-GL} (Fig. 3.2), but all of the signatures are less dramatic than for per-capita release of CO_2 from the combustion of fossil fuels (Fig. 3.1). The exponential rise of pC^{GL} prior to 1910 (Fig. 3.1b) is replaced by a slow, steady, nearly linear rise in pC^{EQ-GL} (Fig. 3.2b) over this same period of time. The time series for pC^{EQ-GL} has a much stronger representation of agriculture than the time series of pC^{GL}. Much of the atmospheric release of CH_4 and N_2O, historically, has been associated with the production of food (Sects. 1.2.3.3 and 1.2.3.4), as has CO_2 released due to land use change. The recent rise in the release of atmospheric CO_2 due to the development of China imposes a different signature when viewed in the context of only fossil fuel CO_2 (start of 3^d growth spurt, Fig. 3.1b) than when examined using the UNFCCC basket of GHGs (moderate uptick, Fig. 3.2b). The major reason for the different appearance, when viewed using these two metrics, is a slower rate of rise of the human release of CH_4 (Fig. 3.3b) during the time when emission of CO_2 from the combustion of fossil fuel from China had accelerated.

The contrast in how per-capita emissions appear, when viewed in terms of release of CO_2 by the combustion of fossil fuels versus release of the UNFCCC basket of GHGs, epitomizes the challenge faced for achievement of the Paris Climate Agreement. The world's peoples must eat. Production of food imposes a considerable burden on atmospheric CH_4 and N_2O, as well as atmospheric CO_2 from the parts of the world that rely on slash and burn agriculture. Whereas future levels of N_2O are projected to rise in both RCP 2.6 (van Vuuren et al. 2011) and RCP 4.5 (Thomson et al. 2011), future levels of CH_4 decline by end of century for both of these RCP scenarios (Fig. 2.1). Reducing the emission of the UNFCCC basket of GHGs will require developing methods to feed a growing global population while, at the same time, reducing emissions of CH_4, N_2O, and CO_2 from land use change. We would be remiss if we did not mention that emission of GHGs could be reduced, particularly the release of CH_4, if more of the world adopted a plant-based diet (Stehfest et al. 2009; Pierrehumbert and Eshel 2015).

3.2.2 *National*

Figure 3.4 shows maps of the emission of CO_2 due to combustion of fossil fuels, flaring, and cement manufacture from individual nations (CO_2^{FF-IN}) for four selected years. Data are based on national inventories maintained and regularly updated by the US CDIAC (Boden et al. 2013), and are shown in units of Gt CO_2 per year. The maps reflect modern political boundaries. The CDIAC estimates are widely used in the climate community and are generally considered to be very reliable (Le Quéré et al. 2015), although there is some debate about the accuracy of the estimates for China in recent years (Guan et al. 2012; Liu et al. 2015). Our maps rely on the most recent CDIAC emission estimates for China, as well as other nations, to ensure a consistent approach for all countries.

Figure 3.5 shows national maps of per-capita release of CO_2 due to the combustion of fossil fuel (pC^{IN}). The population of individual nations is based on data provided by the Population Division of the United Nations (UN) Department of Economic and Social Affairs (see Methods, Fig. 3.1). Values of pC^{IN} are presented in units of metric tons of CO_2 per person per year, abbreviated as t CO_2 ppy.

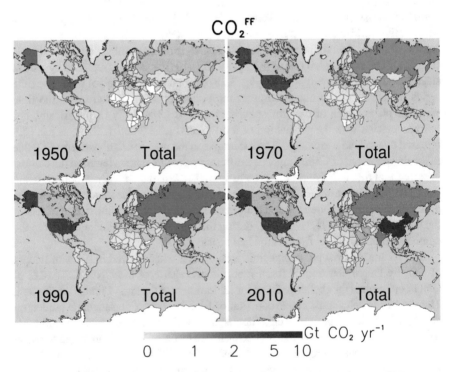

Fig. 3.4 Atmospheric fossil fuel CO_2 emission maps, 1950–2010. Emissions of CO_2^{FF-IN} in units of 10^9 metric tons of CO_2 per year (Gt CO_2 year^{-1}). Maps reflect modern political boundaries. The progression of CO_2^{FF-IN} from 1950 onwards is more informative when viewed as an animation, which can be found at: http://parisbeaconofhope.org/index_animations.htm. See Methods for further information

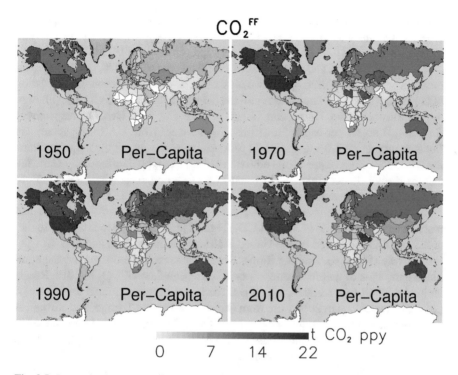

Fig. 3.5 Per-capita fossil fuel CO_2 emission maps, 1950–2010. Per-capita national emissions of CO_2^{FF}, termed pC^{IN}, in units of metric tons of CO_2 per person per year (t CO_2 ppy). The color bar was chosen to highlight emissions from large nations that dominate the global burden of total emissions, and therefore does not cover the full range of pC^{IN}. In 2010, the largest values of pC^{IN} were from Qatar and the nation of Trinidad and Tobago, at 44.7 and 39.5 t CO_2 ppy, respectively. See Methods for further information

The US emitted 2.6 Gt CO_2 in 1950, which was the largest individual national source, followed by the former Soviet Union and the UK, at 0.67 and 0.52 Gt CO_2, respectively (Fig. 3.4). At that time, there was a wide disparity in pC^{IN}. The island nation of Bahrain led the way at 29.5 t CO_2 ppy, followed by Luxembourg and Kuwait at 25.1 and 20.0 t CO_2 ppy, respectively. We have chosen both the color bar for Fig. 3.5 (does not cover the full range of pC^{IN}) and the method of presentation (world map) to highlight major emitters in terms of CO_2^{FF-IN}, rather than small nations that have very large values of pC^{IN}. Of the major emitters in 1950 (i.e., top six emitters in terms of CO_2^{FF-IN}), the US had a pC^{IN} of 16.3 t CO_2 ppy, followed by Canada and the UK, at 11.6 and 10.3 t CO_2 ppy, respectively. In 1950, China emitted 0.081 Gt CO_2, with a per-capita emission of 0.15 t CO_2 ppy.

The release of CO_2 by the combustion of fossil fuels was in the midst of a rapid rise in 1970 (Fig. 3.1). The US was the largest emitter, at 4.38 Gt CO_2, followed by the former Soviet Union and Germany, at 2.32 and 1.04 Gt CO_2, respectively (Fig. 3.4). In 1970, the largest per-capita emissions were from the nations of Qatar, UAE, and Brunei Darussalam, at 69.2, 64.9, and 63.3 t CO_2 ppy, respectively. Of the top six emitters in terms of CO_2^{FF-IN}, the US had the highest per-capita emission at 20.9 t CO_2 ppy, followed by Germany and the UK at 13.3 and 12.2 t CO_2 ppy, respectively (Fig. 3.5). In 1970, China emitted 0.97 Gt CO_2, with a per-capita emission of 0.78 t CO_2 ppy.

The global value of CO_2^{FF} was lower in 1990 compared to 1970, due to improvements in efficiency spurred by the two oil shocks, as well as the economic recession of the 1980s (Fig. 3.1). The US was still the largest emitter, at 4.95 Gt CO_2, followed by the former Soviet Union and China, at 3.72 and 2.50 Gt CO_2, respectively (Fig. 3.4). Largest per-capita emissions in 1990 were from UAE, Singapore, and Luxembourg, at 29.2, 28.8, and 27.2 t CO_2 ppy, respectively. Of the major emitters, the US had the highest per-capita emission at 19.6 t CO_2 ppy, followed by Germany and the former Soviet Union, at 13.1 and 12.9 t CO_2 ppy, respectively (Fig. 3.5). In 1990, the per-capita emission of CO_2^{FF-IN} from China was 2.15 t CO_2 ppy.

In 2010, global emissions of CO_2 due to the combustion of fossil fuels had reached an all-time high of 33.5 Gt CO_2 (Fig. 3.1).[17] China was the largest emitter, at 8.38 Gt CO_2, followed by the United States and India, at 5.56 and 1.97 Gt CO_2, respectively (Fig. 3.4). Had the former Soviet Union remained together, the combined emissions of member nations would have been 2.65 Gt CO_2 in 2010. Russia emitted 1.77 Gt CO_2 in 2010, which was the fourth highest national total. Largest per-capita emissions in 2010 were from Qatar, Trinidad and Tobago, and Kuwait, at 44.7, 39.3, and 31.0 t CO_2 ppy, respectively. Of the top six emitters in 2010, the US still had the highest per-capita emission at 17.9 t CO_2 ppy, followed by Russia and Germany, at 12.3 and 9.7 t CO_2 ppy, respectively (Fig. 3.5). In 2010, the per-capita emission of CO_2^{FF-IN} from India was 1.60 t CO_2 ppy, whereas per-capita emissions from China had risen to 6.22 t CO_2 ppy.

Figure 3.6 shows maps of the emission of $CO_2^{FF} + CO_2^{LUC} + CH_4 + N_2O$, expressed as CO_2-eq, from individual nations (CO_2^{EQ-IN}) for 1990 and 2010. Emission of CH_4 and N_2O from individual nations is based on EDGAR (Rogelj et al. 2014) and emission of CO_2 from land use change is based on data provided by the United Nations Food and Agriculture Organization (FAO) (Houghton et al. 2012). Figure 3.7 shows per-capita emission of $CO_2^{FF} + CO_2^{LUC} + CH_4 + N_2O$ from the world's nations (pC^{EQ-IN}), again for 1990 and 2010. As for Fig. 3.5, the color bar in Fig. 3.7 has been chosen to highlight the major emitters, rather than all nations. And, as noted above, values of CO_2^{LUC} from individual nations are available only from 1990 onwards, so global maps for CO_2^{EQ-IN} cannot be extended as far back in time as for CO_2^{FF-IN}.

Table 3.2 lists the top 12 emitters, in terms of $CO_2^{FF} + CO_2^{LUC} + CH_4 + N_2O$, for 1990 and 2010. The ascension of China, which was third in global emissions in 1990 and top in 2010, is apparent in Fig. 3.6 (national totals), Fig. 3.7 (per-capita), and Table 3.2. Over this two decade period, CO_2^{EQ-IN} from China nearly tripled, and the per-capita emission more than doubled. India, which now ranks third in the world in terms of national value of CO_2-eq emission, saw its emissions double from 1990 to 2010, while the per-capita emissions from this nation only rose by 35 %. As will be apparent in Sect. 3.3, GHG emissions from India are projected to play an increasingly larger role in the global total over the next four decades.

---

[17] In 2014, another all-time high of 35.9 Gt CO_2 was reached. It is likely this annual emission value will be surpassed in both 2015 as well as 2016, once data for these years are released.

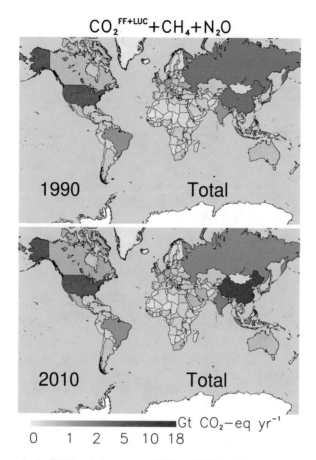

Fig. 3.6 Atmospheric GHG emission maps, 1990 and 2010. National emissions of CO_2^{FF} + CO_2^{LUC} + CH_4 + N_2O in units of 10^9 metric tons of CO_2-eq per year (Gt CO_2-eq year^{-1}). See Methods for further information

Table 3.2 has some additional numbers worth noting. The top 12 emitters contributed 65.3 % of the global total in 1990, and 62.6 % of the global total in 2010. Over this two decade period, global total emission of CO_2-eq rose by 32 %, with nearly no change in global per-capita emissions. Most interestingly, the per-capita emission of the top 12, in aggregate, nearly equaled the global per-capita emission for both 1990 and 2010. In other words, reducing the emission of GHGs to achieve the goal of the Paris Climate Agreement is a global problem: the actions of any one nation, or handful of nations, will have little effect unless the majority of nations participate.

In conclusion of this section, we shall make mention of the numerical entries for Germany in Table 3.2. The pC-eq^{IN} of Germany fell from 15.2 t CO_2 ppy in 1990 to 10.9 t CO_2 ppy in 2010. The drop in per-capita emission of Germany is also apparent in Fig. 3.7. As highlighted towards the end of Chap. 4, Germany has set the standard for generation of energy by renewables that release little or no GHGs, which the rest of the world will have to emulate to achieve the goal of the Paris Climate Agreement.

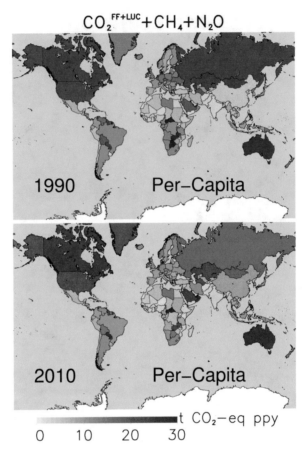

Fig. 3.7 Per-capita atmospheric GHG emission maps, 1990 and 2010. Per-capita national emissions of $CO_2^{FF} + CO_2^{LUC} + CH_4 + N_2O$, termed $pC^{EQ\text{-}IN}$, in units of metric tons of CO_2-eq per person per year (t CO_2-eq ppy). As for Fig. 3.4, the color bar was chosen to highlight emissions from large nations that dominate the global burden of total emissions. In 2010, the largest values of $pC^{EQ\text{-}IN}$ were from Qatar and Trinidad and Tobago, at 75.4 and 54.3 t CO_2-eq ppy, respectively. See Methods for further information

3.3 Future Emissions

Future emissions of GHGs are now examined. As noted in the Introduction, our focus is on emissions of CO_2 due to combustion of fossil fuel[18] and land use change, as well as anthropogenic emissions of CH_4 and N_2O: i.e., the primary drivers of

---

[18] The fossil fuel category also includes emissions from the manufacture of cement and from flaring, which are traditionally lumped into the FF category. Also, all estimates for individual nations or groups of nations include estimates from the combustion of bunker fuels, which is the term used to refer to the mixture of hydrocarbons burned by ships.

Table 3.2 Top Emitters, $CO_2^{FF} + CO_2^{LUC} + CH_{4I} + N_2O$

2010			1990		
Nation	CO_2^{EQ-IN}	pC^{EQ-IN}	Nation	CO_2^{EQ-IN}	pC^{EQ-IN}
China	10.65	7.9	US	5.75	22.8
US	6.15	19.8	USSR	5.38	18.7
India	2.87	2.3	China	3.83	3.30
Russia	2.39	16.7	Brazil	1.65	11.0
Indonesia	2.11	8.7	India	1.48	1.7
Brazil	1.72	8.7	Indonesia	1.42	7.8
Japan	1.13	8.9	Germany	1.20	15.2
Germany	0.88	10.9	Japan	1.17	9.6
Canada	0.85	24.8	UK	0.78	13.6
Iran	0.75	10.2	Canada	0.65	22.3
Mexico	0.66	5.6	France	0.53	9.37
Saudi Arabia	0.64	22.6	Poland	0.53	13.7
Sum, Top 12	30.79	7.82	Sum, Top 12	24.36	7.41
Global	49.19	7.37	Global	37.32	7.53

CO_2^{EQ-IN} in units of Gt CO_2 per year; pC^{EQ-IN} in units of t CO_2 ppy

climate change that are addressed by the Paris Climate Agreement. All emissions are expressed in CO_2-eq, found using GWPs of 28 and 265, respectively, for CH_4 and N_2O (Table 1.1). For the four figures described in this section, total global emissions are shown in all of the top panels. The global emissions from our projections are always represented using grey shading. The thick grey line represents a projection for the UN mid-fertility growth population projection, whereas the top and bottom bounds of the grey shaded region represent emission estimates for high-fertility and low-fertility population projections, respectively. Hence, the grey shaded region represents our estimate of the impact of population on the global emissions of CO_2-eq.

The US, China and India are the top three emitters, nationally, of CO_2-eq (Table 3.2). Therefore, we have chosen to highlight the emission projections from these three nations in the middle and lower panels of the four figures shown in this section. Projections are also shown for Annex I* nations (i.e., all nations *listed* in Table 3.1 other than US), and non-Annex I* nations (i.e., all nations *not listed* in Table 3.1 other than the China and India). We are aware that the Paris Climate Agreement does not make explicit reference to Annex I and non-Annex I nations. Nonetheless, this still seems like a reasonable way to represent the Developed and Developing World, which are referenced in the Paris document.

Figure 3.8 shows projections for the Business as Usual (BAU) scenario. As detailed in Methods, our BAU estimate of global CO_2-eq emissions (grey) considers projections of population from the UN, and forecasts of gross domestic product (GDP) from the Organization for Economic Co-operation and Development (OECD 2016). Data for CO_2-eq emissions from the five groups (US, China, India, Annex I*, and non-Annex I*) from 2000 to 2014 are used to define time series of carbon intensity, I_C, where $I_C = (CO_2\text{-eq emission})/(GDP)$. Past data are used to infer trends in I_C, which are projected forward in time. The world has become more carbon efficient in the past several decades. Not only has $pC^{EQ\text{-}GL}$ fallen from 1990 to 2010 (Table 3.2 and Fig. 3.2b), but world economic output has risen. The BAU projections of CO_2-eq are based on combining forecasts of I_C with forecasts of GDP, an approach known in the climate community as the simplified Kaya Identity (Friedlingstein et al. 2014). A more sophisticated approach, termed the full Kaya Identity, would include additional terms that represent energy demand and energy generation technologies (Raupach et al. 2007). In a sense, we have used the full Kaya Identity approach for Chap. 4, albeit in a global sense.

Figure 3.8a compares our projected global CO_2-eq emissions (grey) to those from RCP 8.5, RCP 4.5, and RCP 2.6. On all of the figures described in this section, our projections and those from RCP represent only $CO_2^{FF} + CO_2^{LUC} + CH_4 + N_2O$, found using the same numerical values of GWP. Figure 3.8a also shows projections of global emissions for the Kyoto basket of GHGs from the Joint Research Center (JRC) of the European Commission (Kitous and Keramidas 2015): their BAU projection, their analysis of the INDCs, and their estimate of the pathway needed to achieve the Paris upper limit of 2 °C warming. The JRC projections for INDCs are for unconditional only (upper orange curve) and unconditional plus conditional (lower orange curve). Finally, the global GHG emission projection for 2030 from the Planbureau voor de Leefomgeving (PBL) Environmental Assessment Agency of the Netherlands, hereafter PBL, is shown for BAU (Admiraal et al. 2015).[19] Figure 3.8b shows the breakdown of global CO_2-eq between the US, China, India, and the rest of the world groups as Annex I* (surrogate for the Developed world) and non-Annex I*.

The BAU projections shown in Fig. 3.8 contain a few important messages. Without any specific attempt to control emission of GHGs, it appears total global emission will fall short of RCP 8.5 by 2030, albeit slightly. In 2060, the BAU projection indicates China and India will be the two top emitters. Not surprisingly, emissions from the Developing World (non-Annex I*) are projected to grow more strongly than for other regions (Fig. 3.8b), even as per-capita emission from the Developing World lags that of other regions (Fig. 3.8c). Our baseline BAU projection for mid-fertility population growth exceeds, by a very small amount, the PBL BAU projection for 2030 (black dot) as well as the JRC BAU projection. However, the grey shaded region of our projection (uncertainty due to population) encompasses the BAU projections from PBL and JRC. Finally, it is evident from the impact of the uncertainty of projected population in 2060 that, while a lower population trajectory is desirable for achievement of the Paris Climate Agreement, more than population control must be implemented. The projected emissions in the decade 2050–2060 for BAU lie about midway between RCP 4.5 and RCP 8.5, which would not enable the goals of Paris to be achieved.

Figure 3.9 shows our projected global emissions of CO_2-eq (grey) for a scenario we call Attain and Hold, Unconditional (AH^{UNC}). For AH^{UNC}, we have assumed emis-

---

[19] PBL has a most informative INDC webpage, at http://infographics.pbl.nl/indc

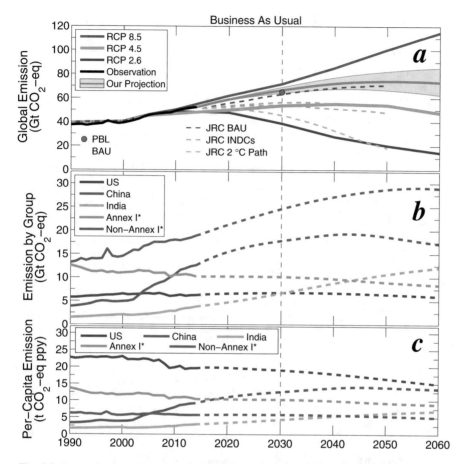

Fig. 3.8 Future GHG projections, Business as Usual (BAU). (**a**) Our projection of global emission of $CO_2^{FF} + CO_2^{LUC} + CH_4 + N_2O$, expressed as CO_2-eq, for a BAU approach; shaded region represents uncertainty based on various population pathways (*grey*). Global emissions of $CO_2^{FF} + CO_2^{LUC} + CH_4 + N_2O$ from RCP 2.6, 4.5, and 8.5, as indicated. Four projections of global emissions for the Kyoto basket of GHGs from JRC (Kitous and Keramidas 2015) are shown: BAU, their analysis of the INDCs, and their estimate of the pathway needed to achieve the Paris upper limit of 2 °C warming. The INDC projections of JRC are for unconditional only (*upper orange curve*) and unconditional plus conditional (*lower orange curve*). Finally, the global GHG emission BAU projection for 2030 from PBL (Admiraal et al. 2015) is shown. (**b**) Our projection of contributions to $CO_2^{FF} + CO_2^{LUC} + CH_4 + N_2O$ from the US, China, India, Annex I*, and non-Annex I*, all for BAU. (**c**) Per-capita emission of $CO_2^{FF} + CO_2^{LUC} + CH_4 + N_2O$ from the five groups, based on our projections in panel (**b**). See Methods for further information

sions follow the submitted INDC, for the 117 nations that have submitted unconditional INDCs to UNFCCC at the time of writing, that include specific quantifiable reductions in GHG emissions. For a few nations, which shall remain unnamed, our best interpretation of their INDC leads to emissions that are larger than we have forecast under BAU. In these instances, the INDC-based forecast is used. Most of the

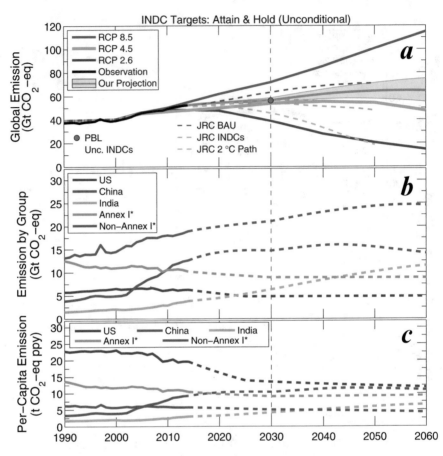

Fig. 3.9 Future GHG projections, Paris Unconditional INDCs, Attain and Hold. Same as Fig. 3.8, except our projections and that of PBL Netherlands (data point at 2030) are for our respective analyses of the Paris INDCs, considering departure from business as usual only for nations that have submitted unconditional INDCs to UNFCCC. For our projections, we assume all unconditional INDCs are followed out to the time of the commitment, and from that point onward carbon emissions hold steady. BAU projections are used for nations that submitted conditional INDCs, and for nations that did not submit an INDC. See Methods for further information

INDCs extend to 2030. The INDC-specific projections of CO_2-eq emissions extend to the target year of each submission.[20] From that year onward, CO_2-eq emissions are assumed to remain constant: hence, the use of "Hold" for this scenario.

Our AH[UNC] projections of CO_2-eq are in extremely close agreement with the unconditional INDC projections of PBL and JRC for year 2030. Our projection tends to run higher than that of JRC for the latter years, most likely because they have assumed continued improvement in carbon intensity for years after 2030. Global emissions remain above RCP 4.5, regardless of population.

---

[20] The majority of the 190 INDCs, about 150, have an end year of 2030. We write "about" because some INDCs have multiple target years, whereas others lack specific target years.

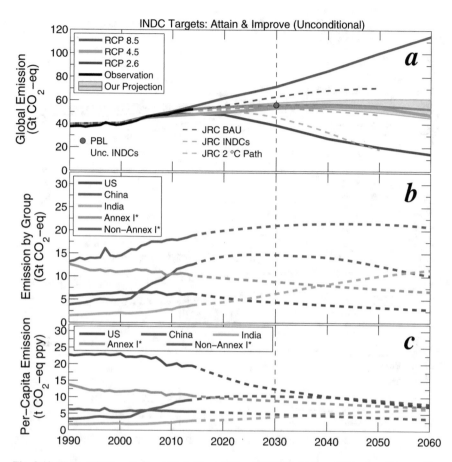

Fig. 3.10 Future GHG projections, Paris Unconditional INDCs, Attain and Improve. Same as Fig. 3.9, except we assume all of the unconditional INDCs are followed out to the time of the commitment and, from that point onward, CO$_2$-eq emissions *continue to decline* at the rate that had been needed for each nation to have achieved its commitment. The data point for PBL Netherlands is the same as that used for Fig. 3.8. See Methods for further information

Figure 3.10 shows our projected global emissions of CO$_2$-eq (grey) for a scenario we call Attain and Improve, Unconditional (AI^{UNC}). For AI^{UNC}, we again consider emissions will follow the specifications of all of the unconditional INDCs that have been submitted to UNFCCC. For this scenario, we assume carbon intensity will continue to improve, after 2030 (or whatever end year was used in the INDC), out to either 2060 or until CO$_2$-eq from a specific nation falls to 50 % of that nation's value in 2030. The projected value of CO$_2$-eq is in extremely close agreement with that of PBL in 2030, and the JRC projection that extends to 2050. The AI^{UNC} global emissions approach those of RCP 4.5 in 2060, but lie above RCP 4.5 for most population projections. Note the strong convergence of per-capita emissions from the US, China, India, and Annex I* in 2060 for AI^{UNC}, towards the value of 7.5 t CO$_2$-eq ppy (Fig. 3.10c). This convergence is in contrast to per-capita emissions from BAU, which exceed this value for the US, China, and Annex I* (Fig. 3.8c).

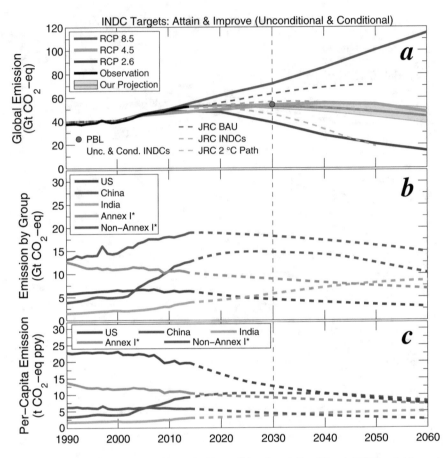

Fig. 3.11 Future GHG projections, Paris Unconditional and Conditional INDCs, Attain and Improve. Same as Fig. 3.10, except our projections consider both unconditional and conditional INDCs. We assume all of the unconditional and conditional INDCs are followed out to the time of the commitment and, from that point onward, CO_2-eq emissions *continue to decline* at the rate that had been needed for each nation to have achieved its commitment. The data point for PBL Netherlands is their projection for 2030, based on the unconditional and conditional INDCs. BAU projections are used for nations that that did not submit an INDC. See Methods for further information

Figure 3.11 shows projections for the final scenario, Attain and Improve, Unconditional and Conditional (AI$^{UNC+COND}$). Here, the treatment has been expanded to consider all 190 nations that have submitted an INDC, i.e., all plans whether conditional or unconditional.[21] Note the extraordinary good agreement with the

---

[21] Some of the INDCs are difficult to interpret quantitatively, with regards to reduction in the emission of GHGs. When in doubt, we used BAU for all projections. For 166 nations, the GHG emmission forecast is based on our best interprepation of the INDC. For the other 24 nations, the forecast is based on BAU, becaue for these nations, the submitted INDC was qualititive rather than quantitative.

PBL projection for 2030, and the JRC time series, both of which consider unconditional and conditional INDCs. Our analysis of the INDCs was conducted in-house, independent of PBL and JRC. A fair amount of judgement was needed to assess some of the plans. We have some trepidation about the veracity of the terms for a few nations (again, unnamed) in the INDC maps shown in Methods. Nonetheless, Fig. 3.11a shows remarkably good agreement between our independent analysis of the INDCs and the estimates of PBL and JRC.

One takeaway from Fig. 3.11 that the Paris Climate Agreement community should embrace is that if the world were to: (a) follow the unconditional and conditional INDCs; (b) commit to continued improvement in carbon intensity out to 2060, then global CO_2-eq emission would likely fall below that of RCP 4.5 regardless of future population. According to our Empirical Model of Global Climate projections, *RCP 4.5 is the 2 °C pathway* (Chap. 2). Of course, as is well known either from this book by now or from the literature (Rogelj et al. 2016), the CMIP5 GCMs indicate a steeper path of CO_2-eq emission reductions is needed to achieve 2 °C. The JRC pathway to achieve 2 °C warming, which is based on these GCMs, is illustrated on the top panel of Figs. 3.8, 3.9, 3.10, and 3.11.

We encourage critical evaluation of our EM-GC approach as well as the GCM forecasts, by other researchers, so that the COP of UNFCCC community has a means to evaluate these starkly contrasting assessments of how steep GHG emission must be reduced, to achieve the goals of the Paris Climate Agreement. In Chap. 4, Implementation, we consider both the RCP 4.5 and RCP 2.6 scenarios.

3.4 Methods

Many of the figures use data from publically available sources. Here, webpage addresses of these archives, citations, and details regarding how data and model output have been processed are provided. Only those figures with "see methods for further information" in the caption are addressed below. Electronic copies and animations of the figures are available on-line at http://parisbeaconofhope.org.

Figure 3.1 shows total global emissions of atmospheric CO_2 from fossil fuels and global population. The CO_2 emissions data were obtained from two files hosted by the Carbon Dioxide Information Analysis Center (CDIAC) at the US Department of Energy's (DOE) Oak Ridge National Laboratory (ORNL):

http://cdiac.ornl.gov/ftp/ndp030/global.1751_2013.ems
http://cdiac.ornl.gov/ftp/Global_Carbon_Project/Global_Carbon_Budget_2015_v1.1.xlsx

The first file was used for CO_2 emissions from 1820 to 2013; the second file was used to obtain data for 2014. The population data shown in Fig. 3.1a and that was used to find pC^{GL} shown in Fig. 3.2 originate from two sources. For years up to 1949, data from the Maddison Project (Bolt and van Zanden 2014) in file:

http://www.ggdc.net/maddison/maddison-project/data/mpd_2013-01.xlsx

were used. For 1950 onward, global population is based on 2015 revision of data assembled by the Population Division of the United Nations Department of Economic and Social Affairs,[22] available on line at:

https://esa.un.org/unpd/wpp/DVD/Files/1_Indicators%20(Standard)/EXCEL_ FILES/1_Population/WPP2015_POP_F01_1_TOTAL_POPULATION_BOTH_ SEXES.XLS

Figure 3.2 shows total global emissions of atmospheric CO_2 due to the combustion of fossil fuels (CO_2^{FF}) and land use change (CO_2^{LUC}), emissions of CH_4 and N_2O expressed as CO_2-equivalent, and global population. The data used for CO_2^{FF} and population are the same as described above for Fig. 3.1. Emissions for CO_2^{LUC}, CH_4, and N_2O are based on Representative Concentration Pathway (RCP) values from files hosted by PICR (Meinshausen et al. 2011) at:

http://www.pik-potsdam.de/~mmalte/rcps/data

Data from file 20THCENTURY_EMISSIONS.DAT were used for years up to 2005, the last year covered in this file. Data from file RCP85_EMISSIONS.DAT were used for 2005–2014, because observed CH_4 over the past decade is closer to CH_4 from the RCP 8.5 scenario than any of the other three RCP scenarios. The RCP emissions for CH_4 are in units of 10^6 metric tons of CH_4 (Mt CH_4) and are converted to the CO_2-eq units used in Fig. 3.2 by multiplying the RCP data by 10^{-3} Gt/Mt $\times$ 28, where 28 is the GWP of CH_4 for a 100-year time horizon (IPCC (2013); see also Table 1.1). The conversion for N_2O requires an extra step. The RCP emissions for N_2O are in units of 10^6 metric tons of N (Mt N). However, the N represents *both* nitrogen atoms in a molecule of N_2O. As such, the conversion is accomplished by multiplying the RCP data by 10^{-3} Gt/Mt $\times$ 265 $\times$ (44/28), where 265 is the GWP of N_2O for a 100-year time horizon (IPCC (2013); see also Table 1.1) and 44/28 is the ratio of the molecular weight of N_2O to the molecular weight of N_2.

Figure 3.3 compares global emissions of CH_4 and N_2O from two databases. The top panel shows results from RCP, based on the same files as described for Fig. 3.2. Figure 3.3b compares emissions of CH_4 from RCP to emissions from version 4.2 FT2012 of the Emissions Database for Global Atmospheric Research (EDGAR) database (Rogelj et al. 2014) from the World Total row of file EDGARv42FT2012_ CH4.xls, found at:

http://edgar.jrc.ec.europa.eu/overview.php?v=42FT2012

Figure 3.3c compares emissions of N_2O from RCP to emissions from EDGAR. The EDGAR time series is based on file EDGARv42FT2012_N2O.xls from the same site, again using the EDGAR World Total entry.

Figure 3.4 shows maps of emissions of CO_2^{FF} from individual nations, termed CO_2^{FF-IN}. Data are from the US CDIAC (Boden et al. 2013) placed on-line at:

http://cdiac.ornl.gov/trends/emis/tre_coun.html

Current political boundaries are used for all four panels, and for all map plots in this chapter. Carbon emission from the former USSR is all that is available prior to 1992. Therefore, for years prior to 1992, former members of the USSR are assigned a value for CO_2^{FF} equal to the product of their fractional contribution to the former

---

[22] https://esa.un.org/unpd/wpp/Publications

USSR sum in 1992, times the total for USSR value for earlier years. The change in political boundaries for the rest of the world (i.e., Czech Republic and Slovakia of the former Czechoslovakia; Bosnia and Herzegovina, Croatia, Macedonia, Montenegro, Serbia, Slovenia of the former Yugoslavia; etc.) was handled in the same manner.

Figure 3.5 shows maps of per-capita emissions of CO_2^{FF} from individual nations. Data for CO_2^{FF-IN} are the same as described in Methods for Fig. 3.4. Population data are from the United Nations Department of Economic and Social Affairs, as described in Methods for Fig. 3.1.

Figure 3.6 shows maps of $CO_2^{FF} + CO_2^{LUC} + CH_4 + N_2O$ from individual nations. Data for CO_2^{FF-IN} are as described for Fig. 3.4. Data for emissions of CH_4 and N_2O for individual nations are from version 4.2 FT2012 of the Emissions Database for Global Atmospheric Research (EDGAR) database (Rogelj et al. 2014), available on-line at:

http://edgar.jrc.ec.europa.eu/overview.php?v=42FT2012

Data for CO_2^{LUC} from individual nations are from the Food and Agriculture Organization (FAO) of the United Nations, available on line at:

http://faostat3.fao.org/download/G2/GL/E

A description of the FAO CO_2^{LUC} data set, and estimates of CO_2 released by LUC from other groups, is given by Houghton et al. (2012). These estimates are available starting in 1990.

Figure 3.7 shows maps of per-capita emissions of $CO_2^{FF} + CO_2^{LUC} + CH_4 + N_2O$ from individual nations. Emission data are the same as for Fig. 3.6, and the population of individual nations is from the United Nations Department of Economic and Social Affairs, as described in Methods for Fig. 3.1.

Figures 3.8, 3.9, 3.10, and **3.11** show projections of emissions of $CO_2^{FF} + CO_2^{LUC} + CH_4 + N_2O$, in CO_2-eq, for business as usual (BAU) (Fig. 3.8) and the three scenarios for the Paris INDCs (Figs. 3.9, 3.10, and 3.11). Each is described below.

Figure 3.8 shows projections of future CO_2-eq emissions for BAU. These projections were found by analyzing the world based on division into five groups: US, China, India, Annex I* nations (all nations *listed* in Table 3.1 other than US), and non-Annex I* (all nations *not listed* in Table 3.1 other than China and India). For each of these groups, carbon intensity (I_C) was calculated over years 2000–2014,[23] where I_C is defined as the quotient of $\Sigma(CO_2^{EQ-IN})$ divided by $\Sigma(GDP)$. This approach is the same as used by Friedlingstein et al. (2014), except our projections use CO_2-eq emissions rather than CO_2^{FF} emissions. Values of GDP were obtained from the OECD (2016) database, on line at:

https://data.oecd.org/gdp/gdp-long-term-forecast.htm

---

[23] The use of 2000–2014 to define trends in I_C is somewhat arbitrary. The use of a much shorter time span introduces noise into the analysis, due to temporary economic fluctuations that are not reflective of decadal time-scale shifts. The use of a much longer time span introduces outdated technology into the analysis. We have chosen 2000 as the start time because this represents an inflection in both the global value of CO_2-eq (Fig. 3.2a) and the global per-capita value of this quantity (Fig. 3.2b). The projections shown in Fig. 3.8 are insensitive to small changes in the start date, particularly if the start year for defining trends in I_C is pushed forward in time by a few years.

In all cases, GDP is based on purchasing power parity in units of 10^{12} 2010 US dollars (USD). In other words, we use carbon emissions and GDP for the US, China, and India, since future carbon emissions from these three nations are highlighted in the figures, whereas we use aggregate sums for carbon emission and GDP for the two other groups. The quantity I_C has units of Gt CO_2-eq $/10^{12}$ USD. For the five groups above, in the order listed, I_C declined at an annual rate of 2.06, 2.48, 2.18, 2.20, and 2.05 % from 2000 to 2014. The world is becoming more carbon efficient.

Figures 3.12 and **3.13** show global maps of BAU projections of the emissions of $CO_2^{FF} + CO_2^{LUC} + CH_4 + N_2O$, in CO_2-eq units, for 2030 and 2060. For the US, China, and India, future carbon emissions for BAU were found by multiplying the OECD projection of GDP by the projection of I_C, where I_C was assumed to decline at a rate of 2.06 % per year for the US. For projections of future CO_2-eq emissions from China and India, I_C was assumed to decline at 2.48 % and 2.18 % per year, respectively. For the rest of the world, BAU projections of CO_2-eq emissions were made using the OECD GDP projection for that group, combined with the rate of decline of I_C from that nations group (2.20 % per year for Annex I*, and 2.05 % per year for non-Annex I*). The specific contribution to future GHG emissions from any nation in the Annex I* or non-Annex I* group, which constitute the data shown in Figs. 3.12 and 3.13, was found from the product of the ratio of that nation's relative contribution to the emission total from the group in year 2014, times the projected future emission from the entire group.

The rest of world (nations other than the US, China, and India) have been combined in this aggregate fashion for numerous reasons. Since the US, China, and India were the top emitters in 2010, and are projected to remain the top emitters out to 2060, it seems appropriate to highlight these three nations in Figs. 3.8, 3.9, 3.10, and 3.11. Also, trends in I_C for some nations are skewed by jumps in CO_2^{LUC} that appear to be unrealistic. The 28 members of the European Union at the time of the Paris meeting submitted a single INDC, further supporting the validity of an aggregate approach. Finally, GDP forecasts are not available for many nations, particularly those in the non-Annex I* list. Hence, the use of a Kaya Identity approach for projecting future emissions involves some aggregation of data.

We recognize the future forecast for BAU from a nation that has already greatly reduced its value of I_C, such as Germany, does not fare well under our aggregate method. In other words, the CO_2-eq values for Germany shown in Figs. 3.12 and 3.13 are likely over-estimates, because Germany has reduced GHG emissions more quickly than the nations with which it has been combined. However, the success of Germany for large scale transition to renewables has been prominently mentioned in Sect. 3.2, and is emphasized in Chap. 4. We present maps in the form of Figs. 3.12 and 3.13, rather than tabular information for individual countries, to let the reader know we have indeed treated all 196 nations and 18 territories (Falkland Islands, Gibraltar, Greenland, Saint Pierre and Miquelon, etc.) of the world in our forecasts, while at the same time emphasizing that our approach is designed to provide realistic forecasts for the world in aggregate rather than for all nations.

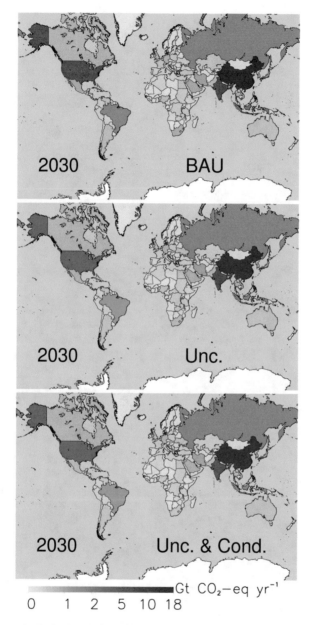

Fig. 3.12 Atmospheric GHG emission maps, Paris INDCs, 2030. National emissions of CO_2^{FF} + CO_2^{LUC} + CH_4 + N_2O in units of 10^9 metric tons of CO_2-eq per year (Gt CO_2-eq year^{-1}), projected to 2030, for the BAU, AH^{UNC}, and $AI^{UNC+COND}$ scenarios

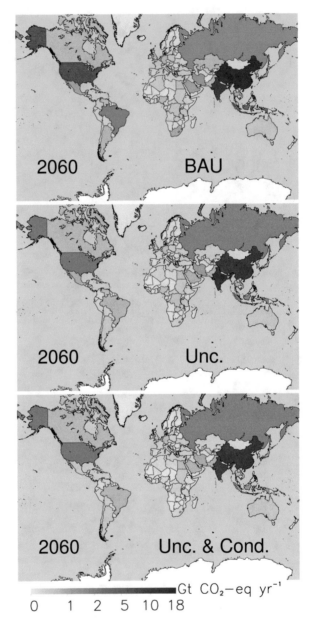

Fig. 3.13 Atmospheric GHG emission maps, Paris INDCs, 2060. Same as Fig. 3.12, but for 2060

Figure 3.8c shows per-capita emissions. Per-capita carbon emissions were found using mid-fertility future population estimates provided by the United Nations Department of Economic and Social Affairs (Methods, Fig. 3.1). The grey shaded region in Fig. 3.8a represents the uncertainty in future values of CO_2-eq, due to population. This was computed by fixing per-capita emission for each nation of the world, at the value calculated using the mid-fertility forecast, then re-computing CO_2-eq emission for either the high-fertility estimate of future population (upper extent of shaded region) or low-fertility estimate (bottom bound).

Figure 3.9 shows projections of CO_2-eq for the Attain and Hold (Unconditional) scenario. Here, future carbon emissions are held at BAU if a country did not submit an INDC, or if the INDC was purely conditional. For the US, the INDC is straight-forward to implement. The last year for which CO_2-eq is available for the US, as for all nations, is 2014. We have assumed CO_2-eq from the US declines by 2.38 %/year, from 2015 to 2025, which leads to a value for CO_2-eq from the US in 2025 that is 27 % below the 2005 value.

The INDC submitted by China focuses solely on emissions of CO_2. Therefore, in all of our projections, we have assumed BAU for emissions of CH_4 and N_2O from China. The INDC from China sets a goal of 60 to 65 % reduction of I_C, relative to the 2005 value, in year 2030. We use 62.5 % in all of our projections. Our imple-mentation of this goal for China, using GDP from OECD, leads to their emissions peaking in year 2026.

The INDC submitted by India has been interpreted to be part unconditional and part conditional. The unconditional component for India reduces I_C by 22.5 % by 2020, relative to 2005, assuming an 8 % per year growth in GDP. The conditional INDC for India imposes additional improvement on I_C, such that by 2030 it is reduced by 35 % of the 2005 value. The proposed additional carbon sink by India is applied to this CO_2 land use term.

It would take more pages than allocated to describe how each and every INDC was handled. Generally, the INDCs fall into three categories. Many give specific emission targets for CO_2-eq, in terms of percentage reduction relative to a base year. For countries that give specific targets, most use a base year of either 1990 or 2005. All European Union nations have based their emission targets off of 1990 values. The preference for 1990 is perhaps a holdover from the Kyoto Protocol. Projections of CO_2-eq emissions for INDCs that have specific targets are straightforward to implement.

Another group of nations have submitted plans to reduce their emission a certain percentage amount, relative to BAU. The implementation of these INDCs is a bit more subjective, as the BAU trajectory must first be calculated. Nonetheless, BAU projections have been found for all nations as outlined above, and the INDC com-mitment then leverages off our BAU projections for this group of INDCs.

A third type of INDC is based on reductions in carbon intensity, or I_C. Evaluation requires calculation of I_C for BAU, which is done as outlined above. There is again some subjectivity, as one must choose which prior years to use for the BAU projec-tion of I_C. And, as noted above, for some nations I_C is particularly difficult to assess, due to large jumps in CO_2^{LUC}. We expect all of these complications will soon be addressed at upcoming meetings of the Conference of the Parties to UNFCCC.

The last detail that must be described is Attain and Hold (AH) versus Attain and Improve (AI). For countries that have submitted specific emission targets for their INDC, such as the US, the emissions under AH are *held fixed* at the targeted value (which for the US is 4.81 Gt CO_2-eq per year, 27 % below the 2005 value) for all years after the specified end year of the INDC (which for the US, is 2025). For countries that submitted carbon emission intensity targets, such as China and India, for all years after the specified end year (which is 2030 for both China and India), the annual decline of carbon intensity is assumed to revert to BAU. For countries that have submitted INDCs that reflect a percentage reduction relative to BAU, under AH the percentage difference between the BAU and INDC values of CO_2-eq is held fixed, for all years after the INDC end year (i.e., we assume emissions from these countries continue to "hold" steady at the same reduction, relative to BAU, for the latter years).

Under the AI projections, national CO_2-eq emissions are extrapolated forward in time, from the end year of the INDC out to 2060. Under AI, for these countries, values of CO_2-eq are linearly extrapolated forward in time, for the out years. For the US we have extrapolation pC^{EQ-IN} from 13.9 t CO_2-eq ppy in 2025, the value achieved under the INDC submitted by the US, to 7.2 t CO_2-eq ppy in year 2060. This target value in 2060 matches the projection of the Annex I* nations, and is slightly less than the value of pC^{EQ-IN} for China in 2060, 8.0 t CO_2-eq ppy. For countries that have submitted INDCs based on carbon intensity, then for years after the end of the INDC under AH, values of I_C found under BAU for the country's group are assumed to replace the state improvement in I_C. In other words, under AH for these countries, we assume the market will control I_C in the latter years. Under AI for the carbon intensity based INDCs, then I_C is allowed to continue to decline, at the annual rate needed to achieve the goal of the INDC, for the years between the end date of the INDC and 2060. Finally, there were a few INDCs that are not easily classified as having either specific targets, being tied to BAU, or leveraging off of carbon intensity. We used our best judgement for how to handle each of these special cases.

The final detail is that in all cases for AI we have set a floor for CO_2-eq from individual nations, such that it can never fall more than 50 % below the value assumed for 2030.[24] The INDCs of some nations commit to much more aggressive reductions in CO_2-eq than those of other nations. Ultimately, it seemed unrealistic to have CO_2-eq from these nations drop more than 50 % below the 2030 value, when other nations had not yet moved their respective needles. Like many of our assumptions, this too is clearly subject to considerable debate.

References

Adams W, Dirlam JB (1966) Big steel, invention, and innovation. Q J Econ 80:167–189

Admiraal A, den Elzen M, Forsell N, Turkovska O, Roelfsema M, van Soest H (2015) Assessing intended nationally determined contributions to the Paris climate agreement—what are the projected global and national emission levels for 2025–2030?

Boden TA, Marland G, Andres RJ (2013) Global, regional, and national fossil-fuel CO_2 emissions. Carbon Dioxide Information Analysis Center, Oak Ridge National Laboratory, U.S. Department of Energy, Oak Ridge, TN, USA. doi:10.3334/CDIAC/00001_V2013

---

[24] This was decided only after considerable internal discussion among the author team. The discussion focused on whether a floor to CO_2-eq should actually be imposed and, if so, what level to use for the floor.

Bolt J, van Zanden JL (2014) The Maddison Project: collaborative research on historical national accounts. Econ Hist Rev 67(3):627–651. doi:10.1111/1468-0289.12032

Canadell JG, Le Quéré C, Raupach MR, Field CB, Buitenhuis ET, Ciais P, Conway TJ, Gillett NP, Houghton RA, Marland G (2007) Contributions to accelerating atmospheric CO_2 growth from economic activity, carbon intensity, and efficiency of natural sinks. Proc Natl Acad Sci 104(47):18866–18870. doi:10.1073/pnas.0702737104

Canty T, Mascioli NR, Smarte MD, Salawitch RJ (2013) An empirical model of global climate— Part 1: a critical evaluation of volcanic cooling. Atmos Chem Phys 13(8):3997–4031. doi:10.5194/acp-13-3997-2013

Falkner R, Stephan H, Vogler J (2010) International climate policy after Copenhagen: towards a 'building blocks' approach. Global Policy 1(3):252–262. doi:10.1111/j.1758-5899.2010.00045.x

Friedlingstein P, Andrew RM, Rogelj J, Peters GP, Canadell JG, Knutti R, Luderer G, Raupach MR, Schaeffer M, van Vuuren DP, Le Quere C (2014) Persistent growth of CO_2 emissions and implications for reaching climate targets. Nat Geosci 7(10):709–715. doi:10.1038/ngeo2248, http://www.nature.com/ngeo/journal/v7/n10/abs/ngeo2248.html#supplementary-information

Guan D, Liu Z, Geng Y, Lindner S, Hubacek K (2012) The gigatonne gap in China's carbon dioxide inventories. Nature Clim Change 2(9):672–675. http://www.nature.com/nclimate/journal/v2/n9/abs/nclimate1560.html#supplementary-information

Hamilton JD (2003) What is an oil shock? J Econ 113(2):363–398, http://dx.doi.org/10.1016/S0304-4076(02)00207-5

Houghton RA, House JI, Pongratz J, van der Werf GR, DeFries RS, Hansen MC, Le Quéré C, Ramankutty N (2012) Carbon emissions from land use and land-cover change. Biogeosciences 9(12):5125–5142. doi:10.5194/bg-9-5125-2012

IPCC (2013) Climate change 2013: the physical science basis. Contribution of working group I to the fifth assessment report of the intergovernmental panel on climate Change. Cambridge, UK and New York, NY, USA

Kirschke S, Bousquet P, Ciais P, Saunois M, Canadell JG, Dlugokencky EJ, Bergamaschi P, Bergmann D, Blake DR, Bruhwiler L, Cameron-Smith P, Castaldi S, Chevallier F, Feng L, Fraser A, Heimann M, Hodson EL, Houweling S, Josse B, Fraser PJ, Krummel PB, Lamarque J-F, Langenfelds RL, Le Quere C, Naik V, O'Doherty S, Palmer PI, Pison I, Plummer D, Poulter B, Prinn RG, Rigby M, Ringeval B, Santini M, Schmidt M, Shindell DT, Simpson IJ, Spahni R, Steele LP, Strode SA, Sudo K, Szopa S, van der Werf GR, Voulgarakis A, van Weele M, Weiss RF, Williams JE, Zeng G (2013) Three decades of global methane sources and sinks. Nat Geosci 6(10):813–823. doi:10.1038/ngeo1955, http://www.nature.com/ngeo/journal/v6/n10/abs/ngeo1955.html#supplementary-information

Kitous A, Keramidas K (2015) Analysis of scenarios integrating the INDCs

Le Quéré C, Moriarty R, Andrew RM, Canadell JG, Sitch S, Korsbakken JI, Friedlingstein P, Peters GP, Andres RJ, Boden TA, Houghton RA, House JI, Keeling RF, Tans P, Arneth A, Bakker DCE, Barbero L, Bopp L, Chang J, Chevallier F, Chini LP, Ciais P, Fader M, Feely RA, Gkritzalis T, Harris I, Hauck J, Ilyina T, Jain AK, Kato E, Kitidis V, Klein Goldewijk K, Koven C, Landschützer P, Lauvset SK, Lefèvre N, Lenton A, Lima ID, Metzl N, Millero F, Munro DR, Murata A, Nabel JEMS, Nakaoka S, Nojiri Y, O'Brien K, Olsen A, Ono T, Pérez FF, Pfeil B, Pierrot D, Poulter B, Rehder G, Rödenbeck C, Saito S, Schuster U, Schwinger J, Séférian R, Steinhoff T, Stocker BD, Sutton AJ, Takahashi T, Tilbrook B, van der Laan-Luijkx IT, van der Werf GR, van Heuven S, Vandemark D, Viovy N, Wiltshire A, Zaehle S, Zeng N (2015) Global carbon budget 2015. Earth Syst Sci Data 7(2):349–396. doi:10.5194/essd-7-349-2015

Liu Z, Guan D, Wei W, Davis SJ, Ciais P, Bai J, Peng S, Zhang Q, Hubacek K, Marland G, Andres RJ, Crawford-Brown D, Lin J, Zhao H, Hong C, Boden TA, Feng K, Peters GP, Xi F, Liu J, Li Y, Zhao Y, Zeng N, He K (2015) Reduced carbon emission estimates from fossil fuel combustion and cement production in China. Nature 524(7565):335–338. doi:10.1038/nature14677, http://www.nature.com/nature/journal/v524/n7565/abs/nature14677.html#supplementary-information

Meinshausen M, Smith SJ, Calvin K, Daniel JS, Kainuma MLT, Lamarque JF, Matsumoto K, Montzka SA, Raper SCB, Riahi K, Thomson A, Velders GJM, Vuuren DPP (2011) The RCP greenhouse gas concentrations and their extensions from 1765 to 2300. Clim Chang 109(1–2):213–241. doi:10.1007/s10584-011-0156-z

OECD (2016) GDP long-term forecast

Pacala S, Socolow R (2004) Stabilization wedges: solving the climate problem for the next 50 years with current technologies. Science 305(5686):968–972. doi:10.1126/science.1100103

Pierrehumbert RT (2014) Short-lived climate pollution. Annu Rev Earth Planet Sci 42(1):341–379. doi:10.1146/annurev-earth-060313-054843

Pierrehumbert RT, Eshel G (2015) Climate impact of beef: an analysis considering multiple time scales and production methods without use of global warming potentials. Environ Res Lett 10(8):085002

Raupach MR, Marland G, Ciais P, Le Quéré C, Canadell JG, Klepper G, Field CB (2007) Global and regional drivers of accelerating CO_2 emissions. Proc Natl Acad Sci 104(24):10288–10293. doi:10.1073/pnas.0700609104

Riahi K, Rao S, Krey V, Cho C, Chirkov V, Fischer G, Kindermann G, Nakicenovic N, Rafaj P (2011) RCP 8.5—a scenario of comparatively high greenhouse gas emissions. Clim Chang 109(1–2):33–57. doi:10.1007/s10584-011-0149-y

Rogelj J, McCollum D, Smith S (2014) The emissions gap report 2014—a UNEP synthesis report: Chapter 2. Nairobi

Rogelj J, den Elzen M, Höhne N, Fransen T, Fekete H, Winkler H, Schaeffer R, Sha F, Riahi K, Meinshausen M (2016) Paris Agreement climate proposals need a boost to keep warming well below 2 °C. Nature 534(7609):631–639. doi:10.1038/nature18307, http://www.nature.com/nature/journal/v534/n7609/abs/nature18307.html#supplementary-information

Stehfest E, Bouwman L, van Vuuren DP, den Elzen MGJ, Eickhout B, Kabat P (2009) Climate benefits of changing diet. Clim Chang 95(1):83–102. doi:10.1007/s10584-008-9534-6

Thomson AM, Calvin KV, Smith SJ, Kyle GP, Volke A, Patel P, Delgado-Arias S, Bond-Lamberty B, Wise MA, Clarke LE, Edmonds JA (2011) RCP4.5: a pathway for stabilization of radiative forcing by 2100. Clim Chang 109(1–2):77–94. doi:10.1007/s10584-011-0151-4

van Vuuren DP, Stehfest E, Elzen MGJ, Kram T, Vliet J, Deetman S, Isaac M, Klein Goldewijk K, Hof A, Mendoza Beltran A, Oostenrijk R, Ruijven B (2011) RCP2.6: exploring the possibility to keep global mean temperature increase below 2 °C. Clim Chang 109(1–2):95–116. doi:10.1007/s10584-011-0152-3

Velders GJ, Andersen SO, Daniel JS, Fahey DW, McFarland M (2007) The importance of the Montreal Protocol in protecting climate. Proc Natl Acad Sci U S A 104(12):4814–4819. doi:10.1073/pnas.0610328104

Velders GJ, Fahey DW, Daniel JS, McFarland M, Andersen SO (2009) The large contribution of projected HFC emissions to future climate forcing. Proc Natl Acad Sci U S A 106(27):10949–10954. doi:10.1073/pnas.0902817106

Victor DG (2001) The collapse of the Kyoto Protocol and the struggle to slow global warming. Princeton University Press, Princeton

Chapter 4
Implementation

**Brian F. Bennett, Austin P. Hope, Ross J. Salawitch,
Walter R. Tribett, and Timothy P. Canty**

Abstract This chapter provides an overview of reductions in the emission of greenhouse gases (GHGs) that will be needed to achieve either the target (1.5 °C warming) or upper limit (2.0 °C warming) of the Paris Climate Agreement. We quantify how much energy must be produced, either by renewables that do not emit significant levels of atmospheric GHGs or via carbon capture and sequestration (CCS) coupled to fossil fuel power plants, to meet forecast **global** energy demand out to 2060. For the Representative Concentration Pathway (RCP) 4.5 GHG emission trajectory to be matched, which is necessary for having a high probability of achieving the Paris target according to our Empirical Model of Global Climate (EM-GC), then the world must transition to production by renewables of 50 % of total global energy by 2060. For the RCP 2.6 GHG emission trajectory to be matched, which is necessary to achieve the Paris upper limit according to general circulation models (GCMs), then 88 % of the energy generated in 2060 must be supplied either by renewables or combustion of fossil fuels coupled to CCS. We also quantify the probability of achieving the Paris target in the EM-GC framework as a function of future CO_2 emissions. Humans can emit only 82, 69, or 45 % of the prior, cumulative emissions of CO_2 to have either a 50, 66, or 95 % probability of achieving the Paris target of 1.5 °C warming. We also quantify the impact of future atmospheric CH_4 on achieving the goals of the Paris Climate Agreement.

Keywords Greenhouse gas emissions • Global energy demand • Renewable energy • Carbon capture and sequestration • Transient climate response to cumulative carbon emissions

4.1   Introduction

Humankind has benefited enormously from the energy provided by the combustion of fossil fuels. The solid form (coal) initially supplied heat and now provides a considerable portion of the world's electricity; the liquid form (petroleum) fuels our vehicles of transportation; and the gaseous form (methane, or natural gas) is used to supply heat, generate electricity, and power transportation vehicles (Fig. 4.1). If you are reading this sentence indoors or electronically, it is probable that the electricity used to power the lights in your room or the screen of your device originated from heat released upon

© The Author(s) 2017
R.J. Salawitch et al., *Paris Climate Agreement: Beacon of Hope*,
Springer Climate, DOI 10.1007/978-3-319-46939-3_4

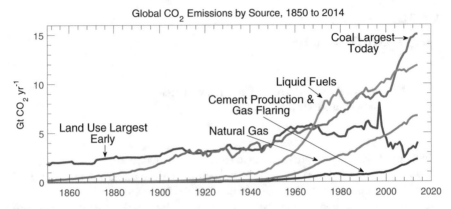

Fig. 4.1 Sources of atmospheric CO_2. Emissions of atmospheric CO_2 from land use change, combustion of solid (coal), liquid (petroleum), and gaseous (methane) forms of fossil fuel, as well as cement manufacturing and flaring. See Methods for further information

combustion of fossil fuel, which generated steam to drive a turbine.[1] The combustion of fossil fuel to power past societies has led to the build-up of atmospheric carbon dioxide (CO_2). Removal of forests due to slash and burn agriculture has also had a considerable effect on atmospheric CO_2 (Fig. 4.1). While it is easy to demonize fossil fuels with statements such as "CO_2 is the greatest waste product of modern society", we must not lose sight of the enormous benefit humankind has gained from the energy supplied upon combustion of the solid, liquid, and gaseous forms of fossil fuels, plus of course the crops grown on land that once used to be forested.[2]

The reliance on fossil fuels to pave our highways, construct and heat our buildings, allow us to visit foreign lands, and power our devices has been driven by two primary factors: the availability of vast reservoirs of readily accessible stocks of this resource and the considerable amount of energy released via the relatively simple combustion process. Alas, the world is now in a bind. Globally averaged atmospheric CO_2, which had a pre-industrial value of 280 parts per million (ppm), has now reached 404 ppm and is rising.[3] Earth's climate has warmed, primarily due to rising atmospheric CO_2.[4]

---

[1] According to analysis of the world's electricity generation by the International Energy Agency, 67.4 % of the world's electricity was provided by the combustion of fossil fuels in 2013. See p. 24 of this summary document: https://www.iea.org/publications/freepublications/publication/KeyWorld_Statistics_2015.pdf

[2] Those quick to judge the developing world for deforestation are urged to consider the dramatic change humans have imposed on landscapes of the developed world, including the US (Bonan 1999).

[3] Readers are encouraged to visit http://www.esrl.noaa.gov/gmd/ccgg/trends/global.html and compare 404 ppm to the value of globally averaged CO_2 recorded by the US National Oceanic and Atmospheric Administration. Odds are CO_2 will be higher, because 404 ppm was measured during late July 2016, when CO_2 is approaching a seasonal low due to the vast NH biosphere that peaks in early fall. A long term, monotonic rise of CO_2 due mainly to the combustion of fossil fuels is imprinted on top of this seasonal variation.

[4] Anyone who questions this statement is invited to read Chap. 1. Overwhelming scientific evidence demonstrates humans are responsible for the rise of CO_2 over the past century and that the increase in global mean surface temperature over this time has been driven primarily by CO_2.

Table 4.1 Atmospheric CO_2 and CH_4 mixing ratios, in parts per million (ppm)

GHG	Present day	2060		2100	
		RCP 4.5	RCP 2.6	RCP 4.5	RCP 2.6
CO_2	404	509	442	538	421
CH_4	1.84	1.80	1.37	1.58	1.25

Unless society is able to *soon* implement provision of electricity, transportation, heat, and industrial energy on a *massive, global scale* that releases little or no GHGs to the atmosphere, we are on a course where the world will experience dire effects of climate change (Lynas 2008).

In this chapter we provide a quantitative analysis of the transformation of energy production that must be put in place for successful implementation of the Paris Climate Agreement. Global warming projections based on the atmospheric, oceanic general circulation models (GCMs) that participated in Climate Model Intercomparison Project Phase 5 (CMIP5) (Taylor et al. 2012) indicate that achieving Paris Climate Agreement upper limit of 2 °C warming will require GHG emissions to follow the Representative Concentration Pathway (RCP) 2.6 trajectory (van Vuuren et al. 2011; Rogelj et al. 2016a). We consider both RCP 2.6 (van Vuuren et al. 2011) and RCP 4.5 (Thomson et al. 2011) emission scenarios in this chapter. Table 4.1 provides present and future atmospheric mixing ratios of CO_2 and CH_4, from both RCP 4.5 and RCP 2.6. Strict reductions in the anthropogenic emission of both GHGs will be needed to achieve either of the RCP 4.5 or RCP 2.6 trajectories.

Much of the focus in this chapter is on emissions of CO_2 because this gas is the primary driver of climate change. We first compare projections of emissions of CO_2 associated with world energy demand developed by the US Energy Information Administration (EIA) to the emissions that will be needed to achieve the RCP 4.5 and RCP 2.6 pathways. We then use satellite observations of light visible from space at night, known as night lights, to illustrate the economic disparity between various parts of the world. For reductions of GHG emissions to occur on a scale to reach RCP 4.5, the *developed* world must transition to a massive use of renewable energy, not only to generate electricity, but also to supply heat and a considerable portion of other energy needs. If the *developing* world is to electrify and industrialize, then this will have to happen in a manner that relies heavily on the use of renewable energy, rather than combustion of fossil fuels, to have a good chance of achieving the upper limit, much less the target, of the Paris Climate Agreement. For the GHG emission reductions of RCP 2.6 to be achieved, carbon capture and sequestration as well as the massive transition to renewables will need to be implemented on a global scale.

We conclude by presenting an analysis of the transient climate response to cumulative CO_2 emissions (TCRE) (Allen et al. 2009; Rogelj et al. 2016b; MacDougall and Friedlingstein 2015), a policy relevant metric highlighted in the Summary for Policy Makers of IPCC (2013). Estimates of TCRE from our EM-GC are compared to values from the CMIP5 GCMs. Finally, we also provide an assessment of the impact of future growth in CH_4, independent of CO_2, on the probability of achieving the Paris Climate Agreement.

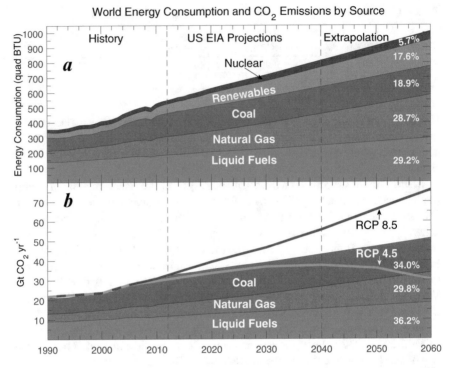

Fig. 4.2 World energy consumption and CO_2 emissions, business as usual. (**a**) Historical (1990–2012) and projected (2012–2040) global energy consumption as a function of fuel source, from the US Energy Information Administration (EIA) and a linear extrapolation of these values out to 2060; (**b**) CO_2 emissions from the coal, natural gas, and liquid fuel components of world energy consumption, provided by EIA for 1990–2040 and extrapolated to 2060, compared to emissions from these sources from RCP 4.5 and RCP 8.5. Numbers to the far right of each wedge are percent of total, for year 2060. See Methods for further information

4.2 World Energy Needs

Increasing demand for energy, as populations expand and standards of living rise, poses a significant challenge to the successful implementation of the Paris Climate Agreement. Figure 4.2a shows a projection of global energy consumption from various sources, in units of 10^{15} British Thermal Units (BTU),[5] provided by the US EIA.[6] This agency forecasts a 70 % increase in world energy consumption in 2040, relative to 2012. We have extrapolated the EIA projections, which stop in 2040, out

---

[5] BTU is a measure of heat, or energy. The conversion of BTU to joule, the unit of energy in the metric system, is usually expressed as 1 BTU = 1055 joule. More information about energy units is at https://www.aps.org/policy/reports/popa-reports/energy/units.cfm

[6] This projection is outlined in a May 2016 report entitled *International Energy Outlook 2016*, available at https://www.eia.gov/forecasts/ieo/pdf/0484(2016).pdf. A concise summary is at https://eos.org/articles/high-energy-growth-fossil-fuel-dependence-forecast-through-2040

to 2060 using linear fits. This extrapolation is conducted because successful implementation of the Paris Climate Agreement will require wholesale transformations in how energy is produced by year 2060.

Figure 4.2b compares the EIA projection of emissions of CO_2 from combustion of coal, natural gas, and liquid fossil fuels needed to meet global energy consumption (colored wedges) to emissions of CO_2 from RCP 4.5 and RCP 8.5 (lines). Again, the EIA estimates extend only to 2040. We have also extrapolated the EIA emission estimates to 2060, for reasons that will soon become apparent. For simplicity, we assume that renewables denote a means of producing energy for which atmospheric release of GHGs is negligible. At the end of this section, potential fallacies of this assumption for renewables are presented. We also assume nuclear energy poses no significant burden to atmospheric GHGs.

Figure 4.2b shows that if the world follows a business as usual approach, emissions of CO_2 from the combustion of fossil fuels will fall between RCP 4.5 and RCP 8.5. The EIA projections are based on forecasts of demand, availability of various technologies, and market forces. There is no attempt to meet any particular climate change goal in this EIA forecast. The good news, we suppose, is that market forces, perhaps combined with environmentalism that acts through the market, appear to be driving the world away from RCP 8.5. The bad news, however, is that the gap between projected emissions of CO_2 and RCP 4.5 is significant in 2040, and grows thereafter.

Figure 4.3 illustrates the dramatic transformation that will have to occur for emissions of CO_2 to follow RCP 4.5 (Thomson et al. 2011) over the 2030–2060 time period. Implicit in the calculations throughout this chapter is the assumption that EIA energy demand projections are met (see Methods for detailed description of how the calculations are conducted). Figure 4.3a shows that for RCP 4.5 CO_2 emissions to be met, the world must place itself on a trajectory whereby half of its energy needs: that is, half of *all* energy used for industry, transportation, heat, electricity, etc., will be realized by renewables that emit little to no GHGs by year 2060. All of the analyses in this chapter extend to 2060 because, quite simply, unless the world soon places itself on this trajectory, it will not be possible to keep global emissions of CO_2 below those of RCP 4.5.

The EIA energy demand forecast is, interestingly, quite similar to that used in the design of RCP 4.5. The four circles in Fig. 4.3a show energy estimates given in Fig. 4.4a of Thomson et al. (2011). While development of energy produced by renewables was an important component of the original RCP 4.5 design, their projection has energy from hydropower, solar, wind, plus geothermal being only ~32 % of the global energy total by end of century.

The reason our projection for the energy share from renewables, 50 % by 2060, differs so much from the RCP 4.5 projection of 32 % by 2100 can be summarized in one word: Fukushima. The Thomson et al. (2011) paper was submitted during September 2010 and was likely in its final stage of review at the time of the Fukushima Daiichi nuclear power plant accident, which occurred during early March 2011. Their RCP 4.5 design included a sizeable slice for growing energy demand to be met by expansion of nuclear energy. Our projections for 2060, on the other hand, rely on extrapolation of the latest EIA projection of nuclear energy from 2030 to 2040. By 2060, nuclear energy is

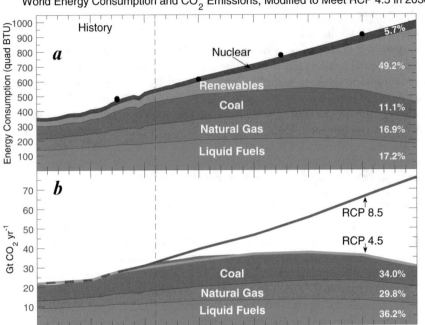

Fig. 4.3 World energy consumption and CO_2 emissions, modified to meet RCP 4.5. Same as Fig. 4.2, except the sum of CO_2 emissions from coal, natural gas, and liquid fossil fuels has been modified to match RCP 4.5 (Thomson et al. 2011) starting in year 2030. For point of comparison, global energy demand used in the design of RCP 4.5 is shown by the four black circles on the top panel. The gap between the EIA-based projection of energy demand and that which can no longer be provided by the combustion of fossil fuels under has been allocated to renewable sources that presumably do not release GHGs. See Methods for further information

projected to supply only 5.7 % of world energy needs.[7] We *do not* allow nuclear energy to grow to accommodate achievement of the RCP 4.5 emissions of CO_2. As such, if global emissions of CO_2 are to meet their RCP 4.5 target, there needs to be a *more rapid transition to renewable energy* than Thomson et al. (2011) had envisioned.

Examination of the present state of affairs for renewables casts this challenge in stark terms. Figure 4.4a shows a breakdown of supply of energy from renewables for the EIA business as usual projection from two categories: combustion of biomass (dark green) and all other sources (light green). Although it is not commonly appreciated, the primary source of energy from renewables throughout the world happens to be combustion of biomass (Kopetz 2013). Much of the developing world heats and cooks using energy derived by burning wood. Energy produced in this manner is classified as renewable, by EIA and others, because the carbon in the

---

[7] EIA bookkeeping has nuclear energy supplying 4.64 % of total, global consumed energy in 2012 and projects supply of 5.64 % in 2040.

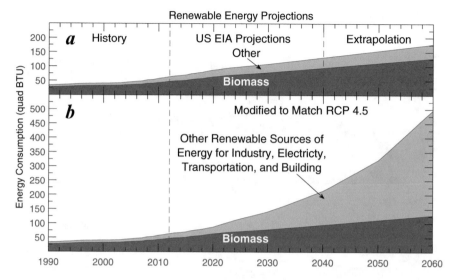

Fig. 4.4 World energy consumption, renewables. (**a**) Historical (1990–2012) and projected (2012–2040) global energy consumption from biomass burning (*bottom wedge*) and other forms of renewables (*top wedge*) from the US EIA and a linear extrapolation of these values out to 2060; (**b**) the biomass burning wedge is the same as used in the top panel. The other forms of renewables wedge shows how much energy must be produced, by forms of renewable energy other than biomass burning (i.e., hydropower, solar, wind, and geothermal), to account for the total amount of renewable energy needed to meet RCP 4.5 in 2060. See Methods for further information

combusted fuel had been in the atmosphere not that long ago.[8] Hydropower is the largest source of electricity from renewables, but total global energy provided by combustion of wood dwarfs that from hydropower.[9] Unfortunately, combustion of wood for heat and cooking in the developing world imposes a serious toll on public health, especially for women and children (Wickramasinghe 2003; Schilmann et al. 2015). Some have proposed expansion of energy from biomass to meet future energy demand (Kopetz 2013). Such an effort will only be tenable if it is conducted in a manner that prevents human exposure to smoke and particulate exhaust. In addition, the generation of energy from biofuels places enormous demand on land use, with the potential to impact food production (Rathmann et al. 2010).

Figure 4.4b shows the extraordinary, rapid growth of the non-biomass forms of renewable energy that will be required in the next four decades to enable emissions of CO_2 to follow RCP 4.5. Our projections are based on the assumption that the EIA energy demand projection will be met. Also, we have forced our biomass forecast to match that of EIA due to the severe harm to public health caused by the present implementation of biomass combustion (Wickramasinghe 2003; Schilmann et al. 2015). Perhaps the growth in energy demand projected by EIA can be dampened due to

---

[8] The salient comparison is trees grow on decadal time scales, whereas the atmospheric origin of the carbon in coal, natural gas, and petroleum is measured on geologic time scales.

[9] Of the myriad of books that describe renewable energy, that one from which we have learned the most is Olah et al. (2009). This book includes extensive sections on hydropower and biomass.

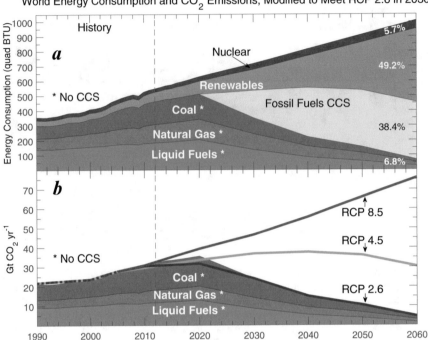

World Energy Consumption and CO_2 Emissions, Modified to Meet RCP 2.6 in 2030

Fig. 4.5 World energy consumption and CO_2 emissions, modified to meet RCP 2.6. Same as Fig. 4.3, except the sum of CO_2 emissions from coal, natural gas, and liquid fossil fuels has been modified to match RCP 2.6 (van Vuuren et al. 2011) starting in year 2030. The time series of consumption of energy produced by renewables has been held fixed at the same value used for the RCP 4.5 projection (Fig. 4.3). The remaining gap has been allocated to carbon capture and sequestration (CCS), a policy option considered by the authors of RCP 2.6. See Methods for further information

improvements in the efficiency of buildings, conservation, electrification of the vehicle fleet,[10] and a decline in population growth. However, it is hard to envision any of these factors dramatically altering the message of Fig. 4.4. Thus, the choice is clear: either the world charts a course towards supplying about half of total, global energy by renewables around year 2060, or CO_2 emissions will exceed those of RCP 4.5.

The majority of the climate modeling community believes that for the 2 °C global warming upper limit of the Paris Climate Agreement to be achieved, CO_2 emissions must be reduced to match those of RCP 2.6 (van Vuuren et al. 2011). As detailed in Chap. 2, forecasts of global warming conducted using our Empirical Model of Global Climate suggest the Paris target will likely be met under RCP 4.5. Regardless, we now extend our forecast to RCP 2.6.

Figure 4.5 illustrates the transformations that will have to occur to match the RCP 2.6 emissions of CO_2 (van Vuuren et al. 2011). The design of RCP 2.6 employed car-

---

[10] Due to the inherent inefficiency of internal combustion on the small scale of a car engine, electric vehicles release considerably less CO_2 per mile traveled than traditional vehicles, even if the electricity used to charge the vehicle's batteries is generated by combustion of fossil fuels.

bon capture and sequestration (CCS) (IPCC 2005), in addition to supply of energy from renewables, to place the world on a low CO_2 emission trajectory. There are many interpretations of CCS (NRC 2015). For our purposes, we will interpret CCS to mean the ability to remove carbon from the exhaust stream of power plants and industrial boilers and then isolate this carbon from the atmosphere, if not permanently then for many centuries. The light blue, orange, and dark blue wedges in Fig. 4.5b represent the energy that can be produced by combustion of fossil fuels that are *not operated* using CCS, in order for the sum of CO_2 emitted from these sources to match RCP 2.6. For illustrative purposes, we have chosen to fix renewables at the same level used in Fig. 4.3a. After all, supplying 50 % of the world's energy by renewables by 2060 is a tall order. The remaining energy deficit is then assigned to CCS. In other words, the gold wedge in Fig. 4.5a represents the amount of energy that must be produced by combustion of fossil fuels with active CCS, to match the RCP 2.6 emissions of CO_2. If the forecasts of global warming by the CMIP5 GCMs are indeed accurate, then for the world to meet the goals of the Paris Climate Agreement, not only will about 50 % of the world's energy need to come from renewables around year 2060, but also about 38 % of global energy must be supplied by combustion of fossil fuels attached to efficient carbon capture and storage. We repeat: by 2060, 50 % of total global energy must be generated by renewables and 38 % must be coupled to efficient CCS to match RCP 2.6 and meet the EIA global energy demand forecast.[11] This is a very tall order.[12]

It is important to emphasize that renewables and CCS are interchangeable for Figs. 4.3 and 4.5. On one hand, CCS can be used to relieve some of the burden assigned to renewables for achievement of RCP 4.5 (Fig. 4.3a), which would be welcome if this technology has matured so that it can be implemented in a safe, efficient, cost effective manner. Alternatively, if by some happenstance renewables are able to capture more than 50 % of the total energy market in 2060, then the need to couple efficient CCS to so much of the world's energy supply to match RCP 2.6 (Fig. 4.5a) would be relieved.

We conclude with sobering thoughts about two technologies that are in the conversation for large-scale production of energy from renewables: hydropower and biofuels. In 2015, hydropower plants generated about 17 % of the world's electricity. Hydropower supplies about 70 % of the total electricity from renewables. The two largest hydropower plants, Three Gorges Dam in China and Itaipú Dam on the border of Brazil and Paraguay, have enormous generating capacities of 22,500 megawatt (MW) and 14,000 MW, respectively. To place these numbers in perspective, a typical coal plant can generate ~700 MW and most nuclear plants are sized at ~1000 MW.

---

[11] Note that Fig. 5 also includes a projection that 5.7 % of the demand in 2060 will be met by nuclear energy. This leaves room for only 6.8 % to be generated by traditional combustion of fossil fuels that is not tied to CCS, in order to meet forecast growth in demand for energy and have GHG emission match RCP 2.6.

[12] Today the world is at 10 % renewables and <1 % CCS. Research efforts on CCS are active throughout the world. In addition to CCS special reports by the Intergovernmental Panel on Climate Change (IPCC 2005) and the US National Academy of Sciences (NRC 2015), the interested reader is directed towards papers such as Hammond and Spargo (2014) and Spigarelli and Kawatra (2013), and references therein.

In some instances, hydropower has a significant GHG burden. At first glance, during operation hydropower should have a negligible GHG burden because electricity is generated from a turbine turned by the force of flowing water. Upon further consideration, there is a GHG burden involved in construction of the power plant, which can be considerable given the size of these massive facilities. A more subtle and much more costly GHG burden is the atmospheric release of CH_4 from decaying biomass in the oxygen deficient flood zone that exists upstream of these massive hydropower facilities. In some cases, particularly the tropics, this creates conditions conducive to release of large amounts of CH_4 from decaying biomass. The GHG burden of a hydropower facility can rival or even exceed that of electricity generation from a comparably sized coal power plant (Fearnside 2002; Gunkel 2009). For massive hydropower plants, there can be little to no climate benefit during the first several decades of operation.

Much has been written about the climate benefit of biofuels and a summary of the debate would take many pages. Numerous books have been written, including *Global Economic and Environmental Aspects of Biofuels* (Pimentel 2012). Of all the renewables, the climate benefit of biofuels is the most controversial.[13] A major point of contention is how the life cycle analysis of biofuels is conducted (Muench and Guenther 2013). In addition to the net benefit for atmospheric CO_2 of biofuels, another concern is atmospheric release of nitrous oxide (N_2O) from intensive application of fertilizer to grow the feedstock (Crutzen et al. 2008). As shown in Chap. 1, N_2O has a global warming potential (GWP) of 265 on a 100-year horizon without consideration of carbon cycle feedback, and a GWP of 298 upon consideration of this feedback (Table 1.1). Since this GHG has an atmospheric lifetime of 121 years, future society would bear the burden for many generations if the atmospheric levels of N_2O were to rise due to aggressive production of biofuels.

4.3   Economic Disparity

Achievement of either the target (1.5 °C) or upper limit (2 °C) of the global warming metrics of the Paris Climate Treaty will require addressing the vast economic disparity that exists in the world. Here, we illustrate this disparity using measurements of night lights obtained by the Visible Infrared Imaging Radiometer Suite (VIIRS) day night band (DNB) radiometer onboard the Suomi National Polar-orbiting Partnership (NPP) platform (Hillger et al. 2013), a joint project of the US National Oceanic and Atmospheric Administration (NOAA) and National Aeronautics and Space Administration (NASA) agencies, as well as gridded population provided by NASA.

Figure 4.6 shows global population and night lights for 2015. Population is from the NASA Socioeconomic Data and Applications Center (SEDAC) Gridded Population of the World version 4 dataset (GPWv4) (Doxsey-Whitfield et al. 2015). Night lights are based on the annual average of cloud free scenes observed by the VIIRS DNB radiometer during 2015. This instrument measures the brightness of

---

[13] We refer those interested in learning more about the debate to this article:
    http://cen.acs.org/articles/85/i51/Costs-Biofuels.html

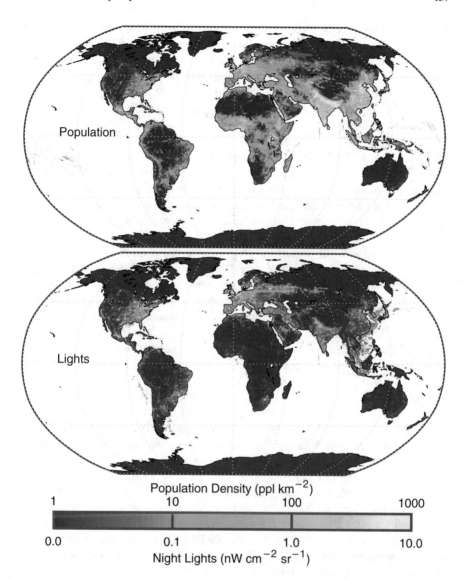

Fig. 4.6 Population and night lights, global. Maps of population and night lights interpolated to the same 0.125° × 0.125° (latitude, longitude) grid. (**a**) Population density for 2015 obtained from the NASA SEDAC GPWv4 dataset; (**b**) night lights measured by the Suomi NPP VIIRS DNB radiometer for cloud free scenes, averaged over all of 2015. The VIIRS night lights data have been processed to remove the dominant effects of aurora borealis and fires. A logarithmic color scale has been used for both population and night lights, to better display the dynamic range of both quantities. Population density is shown in units of people per square kilometer (ppl km^{-2}) and night lights are expressed in units of 10^{-9} watts per square cm per steradian (nW cm^{-2} sr^{-1}). See Methods for further information

light emitted by Earth's surface at wavelengths between 500 and 900 nanometer (nm),[14] at extremely high spatial resolution (Hillger et al. 2013; Liao et al. 2013). Spatial patterns of night lights from VIIRS have been shown to exhibit high correlation with gross domestic product and electricity power consumption (EPC) in China, at multiple spatial scales (Shi et al. 2014).

The VIIRS imagery in Fig. 4.6b has been adjusted so that it better represents economic conditions by removing signals due to the aurora borealis and fires (see Methods). The VIIRS DNB radiometer is sensitive to any light received between 500 and 900 nm; the dominant signals other than EPC are aurora and fires (Liao et al. 2013). Since the aurora generally occur over sparsely populated high latitude regions, a population mask as a function of latitude has been applied that effectively removes the dominant signature from aurora. The influence of wildfires has been removed using NASA Moderate Resolution Imaging Spectroradiometer (MODIS) fire count maps for 2015 (Giglio et al. 2006).

The resulting imagery depicts the geographic distribution of modern infrastructure for electricity as well as population density (Fig. 4.6). The lack of night lights over Africa and much of India, in highly populated regions, is a matter that must be considered by those responsible for implementation of the Paris Climate Agreement. To better illustrate the global disparity in electricity consumption, Fig. 4.7 compares North America with Africa and Fig. 4.8 shows Europe (and parts of eastern Asia) and India (and parts of China). These figures provide dramatic illustration of the haves and the have nots, at least with respect to access to modern infrastructure for electricity.

Figure 4.9 shows scatter plots of night lights versus population for vast regions of the globe. The United States and Europe are lit up at night. The most densely populated regions of China are approaching the night lights density of the US and Europe. While parts of India are starting to become visible from space at night, especially the Haryana and Uttar Pradesh regions that surround Delhi and major cities such as Bengaluru and Hyderabad (Fig. 4.8), the nation as a whole lags the US, Europe, and China, especially the most populated regions (Fig. 4.9). For Africa, most of the night light measurements are below the lowest value shown in the graph. The panel for Africa has fewer lines than the panels for US, Europe, China, and India because only the upper end of the night lights distribution over Africa (95th and 75th percentile, and a single median) are large enough to be displayed on the vertical scale used to display the measurements.

The challenge the world must overcome to slow the emission of GHGs is perhaps best encapsulated by Chinedu Ositadinma Nebo, Minister of Power for The Federal Republic of Nigeria, and Sospeter Muhongo, Energy Minister of the United Republic of Tanzania. When asked during the 2014 US-Africa Leaders' Summit[15]

---

[14] This covers most of the visible spectrum and extends into the near infrared.

[15] The full statements of Minister Nebo and Minister Muhongo are at:
    http://www.scientificamerican.com/article/africa-needs-fossil-fuels-to-end-energy-apartheid

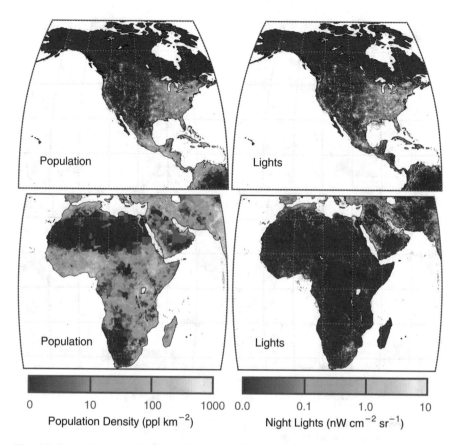

Fig. 4.7 Population and night lights, North America and Africa. Same as Fig. 4.6, except these two regions have been enlarged, for visual clarity

whether their countries would consider development via renewable energy, Minister Nebo stated:

> Africa is hugely in darkness. Whatever we can do to get Africa from a place of darkness to a place of light … I think we should encourage that to happen.

and Minister Muhongo replied:

> We in Africa, we should not be in the discussion of whether we should use coal or not. In my country of Tanzania, we are going to use our natural resources because we have reserves which go beyond 5 billion tons

Implicit in the replies of Ministers Nebo and Muhongo is the notion that for countries such as Nigeria and Tanzania to move their economies forward without relying on domestic reserves of fossil fuel, such that their citizens achieve a standard of living comparable to that of nations on other continents, the developed world must support this effort via payment of a so-called climate rent (Jakob and Hilaire 2015; Bauer et al. 2016). The climate rent is predicated on two notions: (1)

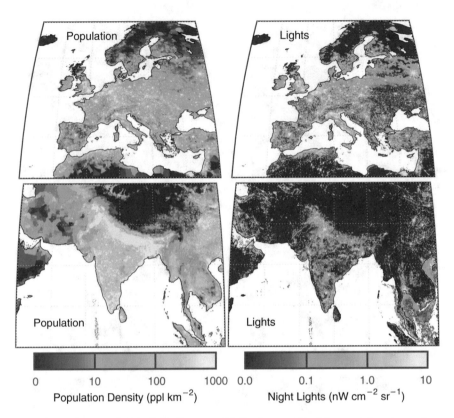

Fig. 4.8 Population and night lights, Europe and India. Same as Fig. 4.6, except these two regions have been enlarged, for visual clarity

the developed world is responsible for conditions that have led to the impending global warming crisis; (2) combustion of fossil fuel remains the most cost effective means of developing an economy, especially if harmful environmental effects of pollutants are not considered. For the developing world to pursue other means of developing their economies, advocates of the climate rent concept would argue this route should be supported, both financially and via technology, by the developed world. Only an agreement perceived to be equitable by all participants can result in the strict limits on the future use of fossil fuel that will be needed to avoid dire effects of global warming.

The Paris Climate Agreement will utilize the Green Climate Fund (GCF),[16] which was established in 2010 to assist developing countries improve their standard of living in a climate friendly manner.[17] As of time of writing, the GCF had raised

---

[16] The resources of the Green Climate Fund can be followed at:

http://www.greenclimate.fund/partners/contributors/resource-mobilization

[17] The Paris Climate Agreement mentions two additional sources to assist the transition, the Least Developed Countries Fund and the Special Climate Change Fund. Neither are nearly as well

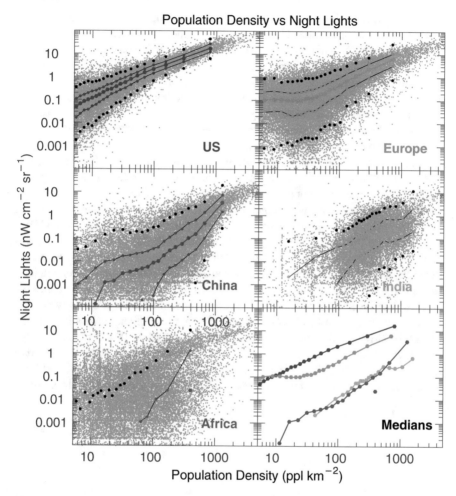

Fig. 4.9 Scatter plots of night lights versus population. Panels show data for the United States, Europe, China, India, and Africa for 2015. Speckles are individual VIIRS night lights versus population density, from the 0.125° × 0.125° grid of the specific country or continent. The *largest colored circles* for each panel show the median values of night lights and population, after the data has been grouped into 20 bins. The *smaller colored circles* show the 25th and 75th percentile of the data in these bins; the black circles show the 5th and 95th percentile. The last panel compares medians for the five geographic regions. See Methods for further information

$10.2 billion USD from 42 governments The GCF goal is to collect and disburse $100 billion USD per year by 2020. While it would be foolish to sneer at such a sincere effort, and who among us can ever envision writing a check for $100 billion,

---

funded as the GCF, at present. Also, the GCF is mentioned in a more prominent fashion.

much less $10.2 billion, we can't help but point out that the total population of the non-Annex I nations in the United Nations Framework Convention on Climate Change (UNFCCC) that have submitted conditional Intended Nationally Determined Contributions (INDCs)[18] is approximately 4 billion people. If fully funded at the $100 billion USD goal, the GCF would contribute only $25 per person per year to assist the developing world transform energy production to avert the adverse effects of climate change. While we can only speculate, it is likely that financial assistance at a higher level would be needed to assuage Energy Ministers such as Chinedu Ositadinma Nebo and Sospeter Muhongo.

We conclude this section by highlighting two of the successful efforts for electrification in Africa presently taking place that involve installation of solar power plants. Large-scale solar photovoltaic installations are being funded by a company named Gigawatt Global[19] whose mission is to invest in the provision of renewable energy to Africa and other under-served, emerging markets. An 8.5 megawatt (MW), grid connected solar photovoltaic system consisting of 28,360 arrays has been operational in Rwanda since September 2014. Other projects are being developed in the Republic of Burundi and Nigeria. Each of these projects consists of a power purchase agreement (PPA) that returns revenue to the consortium of investors who finance the purchase and installation of the system. A company named Solar Reserve, which has successfully installed a 110 MW concentrated solar power plant in Nevada, is developing a 100 MW facility due to open in 2018 in South Africa.[20] Successful renewable energy ventures in Africa such as those financed by Gigawatt Global and Solar Reserve provide hope that Africa, India, and the rest of the developing world can indeed manage to electrify by some means other than the combustion of fossil fuel.

---

[18] This population includes the people of India and all African nations that have participated, but does not include China, since by most interpretations the INDC submitted by China is unconditional. The distinction between conditional and unconditional INDCs is given in Chap. 3. Finally, we are aware that the Paris Climate Agreement does not make explicit mention of Annex I and non-Annex I nations. Nonetheless, non-Annex I is UNFCCC terminology. Furthermore, the list of non-Annex I nations that have submitted purely conditional INDCs corresponds to a roster of countries most would say provides a reasonable representation of the developing world.

[19] Interested readers can learn more about Gigawatt Global at http://gigawattglobal.com

[20] The Solar Reserve, Crescent Dunes Solar Energy Facility in Nevada is described at:

http://www.scientificamerican.com/article/new-concentrating-solar-tower-is-worth-its-salt-with-24-7-power

and the Solar Reserve project in South Africa is described at:

http://www.solarreserve.com/en/global-projects/csp/redstone.

These projects are based on concentrated solar, which operates on the principle of collecting sunlight with reflectors to generate heat that produces steam, which is then used to produce electricity via traditional turbine technology.

4.4 Emission Metrics

In this section, the transient climate response to cumulative CO_2 emissions (TCRE) metric highlighted in IPCC (2013) is described, in terms of the CMIP5 GCMs and our Empirical Model of Global Climate (EM-GC) (Canty et al. 2013). The TCRE metric relates the rise in global mean surface temperature (GMST) to the cumulative amount of anthropogenic carbon released to the atmosphere, by all sources. According to IPCC (2013), the likely range for TCRE is 0.8 to 2.5 °C warming relative to pre-industrial baseline, per 1000 Gt C of CO_2 emissions.[21]

The sensitivity of global warming forecasts using our EM-GC framework to the future atmospheric levels of CH_4 is also examined. This is especially important because a number of nations, including the US, are planning to fulfill their Paris INDC commitment by producing increasingly large percentages of electricity by the combustion of methane, rather than coal. Combustion of CH_4 yields about 70 % more energy than combustion of coal, per molecule of CO_2 released. Hence, the transition from coal to natural gas is touted by many as being climate friendly. However, if only a small percentage of CH_4 escapes to the atmosphere at any stage prior to combustion, then the switch to natural gas can exert a climate penalty due to the large GWP of CH_4 (Howarth et al. 2011).

4.4.1 CO_2

Figure 4.10 compares estimates of TCRE from the CMIP5 GCMs (Taylor et al. 2012) and our EM-GC (Canty et al. 2013). The CMIP5 GCM points shown on both panels are the same, and are taken from Figs. SPM.10 and TFE.8 of IPCC (2013). The figure shows the rise in global mean surface temperature (GMST) relative to a pre-industrial baseline (ΔT). Here, years 1861–1880 are used to define the pre-industrial baseline so that our TCRE figures are as close as possible to the representation in IPCC (2013). The observed value of ΔT for the time period 2006–2015 from the Climatic Research Unit (CRU) of the University of East Anglia data record (Jones et al. 2012) is 0.808 °C upon use of the 1861–1880 baseline.[22]

The values of ΔT shown in Fig. 4.10 are based on EM-GC simulations using GHG and aerosol precursor emissions from RCP 2.6 (van Vuuren et al. 2011), RCP

---

[21] Recall that Gt, the abbreviation for giga ton, refers to 10^9 metric tons of carbon.

[22] As noted at the start of Chap. 2, CRU-based ΔT = 0.828 °C for 2006–2015 if a baseline of 1850–1900 is used. The 50 year baseline has been used to represent pre-industrial in all other sections of this book. Various baselines are used in IPCC (2013), which makes quantitative comparison of some of the figures a bit of a challenge. The difference in ΔT found using these two baseline periods, 0.02 °C, is 1 % of the Paris upper limit of 2 °C warming and, as such, is inconsequential. Nonetheless, we conduct the TCRE analysis in the same manner as IPCC (2013) to avoid criticism for using a different baseline. Had we used 1861–1880 for the baseline period throughout the book, numerical values of ΔT would have been 0.02 °C smaller than shown.

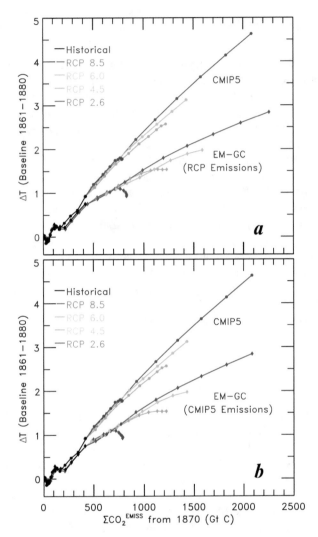

Fig. 4.10 Transient climate response to cumulative CO_2 emissions, in units of Gt C. Both panels show rise in GMST relative to an 1861–1880 baseline (ΔT) from CMIP5 GCMs as a function of cumulative CO_2 emissions for RCP 2.6, 4.5, 6.0, and 8.5. (**a**) Rise in GMST found using our Empirical Model of Global Climate (EM-GC) for the four RCP scenarios, run using the IPCC (2013) best estimate for ΔRF due to tropospheric aerosols between 1750 and 2011 of -0.9 W m^{-2} and OHC based on the average of six data records shown in Fig. 2.8. The computed cumulative CO_2 emissions are based on our summation of data archived in files that drove the various RCP scenarios; (**b**) same as (**a**), except the rise in GMST from our EM-GC is displayed as a function of the cumulative CO_2 emissions associated with the CMIP5 GCMs, which are lower than the CO_2 emissions that drove the RCP scenarios. See Methods for further information

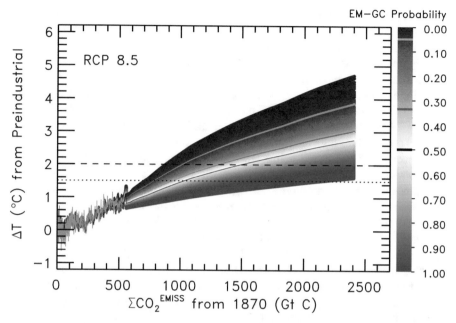

Fig. 4.11 Transient climate response to cumulative CO_2 emissions, RCP 8.5. Simulations of the rise in GMST relative to an 1861–1880 baseline (ΔT) found using our EM-GC plotted versus cumulative CO_2 emission, in units of Gt C. Paris Climate Agreement target and upper limit of 1.5 and 2.0 °C warming are denoted by the *dotted* and *dashed lines*, respectively. The EM-GC projections (*red, white,* and *blue colors*) represent the probability that the future value of ΔT will rise to the indicated level, considering only acceptable fits to the climate record (i.e., $\chi^2 \leq 2$). The light grey, dark grey, and black curves represent the 95, 66, and 50 % probabilities of either the Paris target (intersection of *dotted horizontal lines* with the respective curve) or upper limit (intersection of *dashed line* with curves) being achieved (see text). See Methods as well as Chap. 2 for further information

4.5 (Thomson et al. 2011), RCP 6.0 (Masui et al. 2011), and RCP 8.5 (Riahi et al. 2011). Each of these four runs used the IPCC (2013) best estimate for radiative forcing of climate due to tropospheric aerosols of −0.9 W m^{-2} in year 2011 (AerRF$_{2011}$), as well as ocean heat content (OHC) based on the average of six data records shown in Fig. 2.8. As discussed in Chap. 2, projections of ΔT are sensitive to AerRF$_{2011}$ (Fig. 2.9) and insensitive to OHC (Fig. 2.10).

Values of ΔT from the CMIP GCMs and our EM-GC shown in Fig. 4.10 are displayed as a function of cumulative CO_2 emissions from land use, fossil fuel combustion, cement manufacturing, and flaring since 1870 (ΣCO_2^{EMISS}). Prior to this point, we have displayed CO_2 emissions using units of Gt CO_2, because this is most appropriate for the Paris Climate Agreement. In Figs. 4.10 and 4.11, however, ΣCO_2^{EMISS} is displayed using Gt C because most of the discussion of TCRE in the peer-reviewed literature (Allen et al. 2009; Rogelj et al. 2016b) and in IPCC (2013) uses Gt C. The distinction between these two units is described by footnote 13 of Chap. 1.

Values of ΣCO_2^{EMISS} computed from the RCP database exceed the amounts of ΣCO_2^{EMISS} displayed in Fig. SPM.10 and TFE.8 of IPCC (2013).[23] This overestimate is due to the use of a GCM with an interactive carbon cycle component for the figures shown in IPCC (2013) that differs from the treatment of the carbon cycle used to drive each of the four RCP specifications. Figure 4.10a shows ΔT found using our EM-GC framework, plotted against ΣCO_2^{EMISS} from the RCP database; Fig. 4.10b shows ΔT from EM-GC plotted against ΣCO_2^{EMISS} taken from the IPCC (2013) TCRE figures. The difference is small, but noticeable, and represents the impact on TCRE of how the interactive carbon cycle is treated.

Values of TCRE found using the EM-GC framework have important policy implications. Figure 4.10 shows that the amount of carbon that can be released into the atmosphere before reaching a particular temperature threshold is estimated to be significantly larger based on calculations using our EM-GC than computed using the CMIP5 GCMs. This result is expected based on different characteristics of these two approaches for modeling GMST that were quantified in Chap. 2. There, we had shown the CMIP5 GCMs tend to simulate a warming of the global climate, over the 1979–2010 time period, which is about a factor of two faster than observations indicate the actual climate system warmed (Fig. 2.13). We also showed that the equilibrium climate sensitivity of the actual climate system is likely to be considerably smaller than that represented by GCMs (Fig. 2.11). The EM-GC projection that larger values of carbon can be emitted before a particular temperature threshold is crossed, compared to the CMIP5 GCM forecasts, is consistent with the emergent understanding that the majority of these GCMs simulate warming rates that are likely too fast.

We pursue the policy impact of temperature thresholds using probabilistic forecasts of global warming. The degeneracy of the climate system, outlined in Chap. 2, is fully considered.[24] Figure 4.11 shows the transient climate response to ΣCO_2^{EMISS} found using the EM-GC framework, constrained by RCP 8.5 emissions. The RCP 8.5 scenario is used for Fig. 4.11 because warming of 2 °C is not exceeded, prior to 2060, for any of the other RCP scenarios in the EM-GC framework. All simulations use OHC based on the average of six data records (Fig. 2.8), and have been weighted by $1/\chi^2$ prior to calculation of the probabilities (Sect. 2.5). This figure shows the

---

[23] This difference can be seen in Fig. 4.10a by comparing the red circle with highest ΔT (CMIP5 GCM value, year 2100) to the red diamond with highest ΔT (EM-GC value, year 2100). Not only is ΔT from the EM-GC lower than ΔT from the CMIP5 GCMs, but it is also associated with a larger value of ΣCO_2^{EMISS}.

[24] Briefly, degeneracy of the climate system refers to the fact that the prior ΔT record can be fit *nearly equally well* assuming large climate feedback and strong aerosol cooling, or weak climate feedback and little to no aerosol cooling. Regardless of what had happened in the past, the radiative impact of aerosols will be diminishing in the future, due to public health concerns that are leading to steep reductions in the emission of aerosol precursors. If we assume the climate feedback inferred from the climate record will persist into the future, then projections of ΔT found using the large climate feedback simulation will exceed those found using weak feedback. The community that studies radiative effects of aerosols is not close to a consensus on which of these two scenarios is more likely to be correct.

Table 4.2 Total cumulative carbon emission that will lead to crossing Paris ΔT thresholds

Warming	Total ΣCO_2^{EMISS}			
	CMIP5 GCMs, 50 %	EM-GC, 95 %	EM-GC, 66 %	EM-GC, 50 %
1.5 °C	633 Gt C	797 Gt C	930 Gt C	1002 Gt C
2.0 °C	842 Gt C	1010 Gt C	1300 Gt C	1480 Gt C

probability that future ΔT will rise to a particular value: i.e., the color bar indicates probabilities and the placement of the color on the chart is at the associated time (horizontal axis) and temperature (vertical axis).[25] Horizontal lines on Fig. 4.11 are drawn at the Paris target (1.5 °C; dotted line) and upper limit (2.0 °C; dashed line). The light grey, dark grey, and black curves represent the 95, 66, and 50 % probabilities that ΔT will remain below a particular value.[26]

Table 4.2 quantifies the cumulative emission of CO_2 that will lead to the Paris target (1.5 °C) or upper limit (2.0 °C) being crossed. Values of ΣCO_2^{EMISS} from the GCMs in Table 4.2 are based on IPCC (2013)[27] and the crossing of the two temperature thresholds is assigned a probability of 50 %, since these GCM projections represent the average forecast of numerous simulations from many models. Those interested in a more detailed probabilistic representation of TCRE from CMIP5 GCMs are referred to Rogelj et al. (2016b). The values of ΣCO_2^{EMISS} used for the EM-GC ΔT forecasts are based on CO_2 emissions used to drive RCP 8.5 (Riahi et al. 2011). For the EM-GC calculations, estimates of ΣCO_2^{EMISS} that would cause global warming to stay below indicated thresholds are given for three probabilities: 95, 66, and 50 %. In other words, if cumulative carbon emission stays below 797 Gt C, then according to our EM-GC forecasts there is a 95 % probability the Paris target of limiting global warming to 1.5 °C will be achieved.

Table 4.2 shows that the CMIP5 GCMs, interpreted literally, place much tighter constraints on how much CO_2 can be released prior to crossing the Paris thresholds of 1.5 and 2.0 °C warming. The value of ΣCO_2^{EMISS} from 1870 to 2014, based on a simple summation of the terms in Fig. 4.1, is 551 Gt C. There is a ~15 % uncertainty

---

[25] The use of the color bar to show probabilistic projections of ΔT is explained in much greater detail in Chap. 2. Briefly, the pure white region of Fig. 4.11a is the most probably outcome for forecast ΔT using RCP 8.5, assuming climate feedback and ocean heat export inferred from the climate record persist into the future. The dark blue region shows plausible but unlikely projections of modest warming and the dark red shows plausible but unlikely projections of strong warming. Probabilities associated with modest warming are close to unity (i.e., it is nearly certain the climate system will warm at least this much) and those associated with strong warming are close to zero (i.e., it is unlikely the climate system will warm to this extent).

[26] As explained in prior footnote, dark red colors represent plausible but unlikely values of strong warming. Since it is unlikely climate will warm to this extent, the dark red color is associated with the low probability of 0.05. The light grey line connects all model outcomes probabilities of 0.05. Since there is only a 5 % chance it will warm this much, there is a 95 % chance that warming will fall below the grey line.

[27] More specifically, these values originate from Fig. SPM.10 and TFE.8 of IPCC (2013).

Table 4.3 Future cumulative carbon emission that will lead to crossing Paris ΔT thresholds

| Warming | Future ΣCO_2^{EMISS} | | | |
	CMIP5 GCMs, 50 %	EM-GC, 95 %	EM-GC, 66 %	EM-GC, 50 %
1.5 °C	82 Gt C	246 Gt C	379 Gt C	451 Gt C
2.0 °C	291 Gt C	459 Gt C	749 Gt C	944 Gt C
	% of past CO_2 emissions that lead to threshold being crossed			
1.5 °C	14.9 %	44.6 %	68.8 %	81.9 %
2.0 °C	52.8 %	83.3 %	136 %	171 %

on ΣCO_2^{EMISS}, driven by the land use change component (e.g., error bar on cumulative emission estimate shown in Fig. TFE.8 of IPCC (2013)).

Numerical estimates of cumulative carbon emission that will lead to the Paris Climate Agreement thresholds being surpassed may serve as an important guide to policy. Table 4.3 shows the future cumulative amount of CO_2 that can be released before a particular threshold is crossed, computed by subtracting 551 Gt C from the entries in Table 4.2. The last two rows of Table 4.3 show the ratio of future cumulative carbon that can be released, divided by 551 Gt C and expressed as percent. In other words, according to Table 4.2, the EM-GC projection indicates there is a 95 % probability of limiting future warming to 2 °C relative to pre-industrial if ΣCO_2^{EMISS} can be restricted to 1010 Gt C. As of 2014, 551 Gt C of CO_2 had been released. Therefore, the remaining amount that can be released is 459 Gt C, or 83.3 % of the prior release.[28] According to the EM-GC forecast of global warming, humans can only emit 45%, 69 %, or 82 % of the prior, cumulative emission of carbon to have a 95 %, 66 %, or 50 % probability, respectively, of achieving the Paris target of 1.5 °C warming. The CMIP5 GCM forecast places a much tighter constraint on the additional release of carbon before the Paris thresholds are breached. For instance, the GCMs project there will be a 50 % probability that warming will exceed 1.5 °C if humans emit only 15 % of prior, cumulative past carbon emissions.

The CMIP5 GCM based values of ΣCO_2^{EMISS} associated with crossing the Paris target seem implausibly small. As stated at the start of this section, the observed rise in ΔT over the decade 2006–2015 is 0.808 °C.[29] The climate modeling community has drawn attention to the apparent linearity between ΔT and ΣCO_2^{EMISS}, particularly for the first 1000 Gt of carbon emission (MacDougall and Friedlingstein 2015). The value of ΣCO_2^{EMISS} up to end of 2010, the mid-point of the 2006–2015 time period, is 508 Gt C. If ΣCO_2^{EMISS} of 508 Gt C has been associated with 0.808 °C warming, and if the relation is truly linear, then the 1.5 °C threshold should be

---

[28] 459 Gt C = 1010 − 551 Gt C; 83.3 % = 100 × (459 Gt C/551 Gt C).

[29] Estimate of observed ΔT on a decadal average is a simple, time-honored way to remove year to year fluctuations in temperature caused by natural variability. We expect some to criticize our approach using temperature data for only 2015 and 2016. However, as shown in Fig. 2.9, the recent El Niño Southern Oscillation event is responsible for values of ΔT being unusually large in the past 12 months.

crossed when ΣCO_2^{EMISS} reaches 943 Gt C. This back of the envelope calculation is close only to the EM-GC values of ΣCO_2^{EMISS} given in Table 4.2 for 50 % and 66 % probability.[30]

Indeed, we can use another line of reasoning to suggest the CMIP5 GCM based values of ΣCO_2^{EMISS} associated with crossing the Paris target are too low. As noted in the introduction to this section, IPCC (2013) stated the likely range for TCRE is 0.8–2.5 °C warming per 1000 Gt C of CO_2 emissions. Our probabilistic projection of ΔT shown in Fig. 4.11, for the point where $\Sigma CO_2^{EMISS} = 1000$ Gt C, is bounded by 0.8 and 2.4 °C, in near perfect agreement with the range stated by (IPCC 2013). Conversely, the CMIP5 GCM estimate that the 1.5 °C threshold will be crossed when $\Sigma CO_2^{EMISS} = 633$ Gt CO_2 implies a warming of 2.4 °C per 1000 Gt C. Simulations conducted in the EM-GC framework suggest this value is possible but highly unlikely.

Science is driven by reproducibility of results. As stated at the end of Chap. 2, we urge that more effort be devoted to assessing GCM-based forecasts of global warming using energy balance approaches such as our EM-GC framework. It is our sincere hope that others will evaluate and publish values of ΣCO_2^{EMISS} and $\Sigma CO_2\text{-eq}^{EMISS}$, such as those in Tables 4.2 and 4.3, using various model frameworks. Time will tell whether our estimates of ΣCO_2^{EMISS} and $\Sigma CO_2\text{-eq}^{EMISS}$ survive the scrutiny of others. In the interim, we urge policy makers to tentatively consider that achieving the target of the Paris Climate Agreement, via the existing INDC pledges, may indeed be a realistic goal.

4.4.2 CH₄

One final complication must be addressed: the potential rise of atmospheric CH_4. The present globally averaged mixing ratio of CH_4, the second most important anthropogenic GHG, is 1.84 ppm.[31] Projected future values of CH_4 diverge by an enormous amount among the four RCP scenarios (Fig. 2.1).

The RCP projections of CH_4 reflect the large uncertainty in future emissions. The RCP 2.6 scenario (van Vuuren et al. 2011) projects a CH_4 mixing ratio of 1.37 ppm in 2060 (Table 4.1; see also Fig. 2.1). Atmospheric CH_4 has numerous human-related sources (Fig. 1.9). The RCP 2.6 design projects a 26 % decline of CH_4 by 2060, due to stringent controls on human release from all sources other than agriculture. Their projection considers the climate benefit of diet, particularly global con-

---

[30] The fact this back of the envelope estimate for ΣCO_2^{EMISS} lies closer to our 66 % probability value for keeping warming below the 1.5 °C, rather than the 50 % outcome, is due to the small non-linearity in ΔT versus ΣCO_2^{EMISS} manifest in the EM-GC framework that is shown in Fig. 4.10a.

[31] Those keeping score are encouraged to visit http://www.esrl.noaa.gov/gmd/ccgg/trends_ch4; this site continually updates the global mean CH_4, albeit with a delay of a few months.

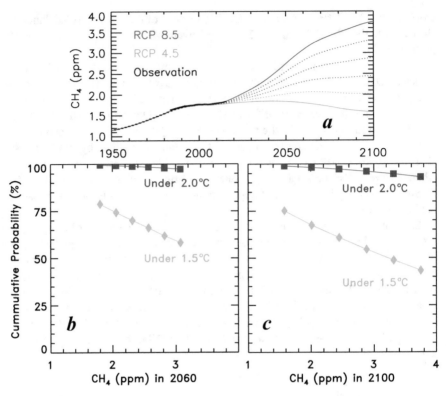

Fig. 4.12 Impact of CH$_4$ on EM-GC projections using RCP 4.5. (**a**) Time series of atmospheric CH$_4$ from observations (Dlugokencky et al. 2009), RCP 4.5 (*blue*, lowest curve) (Thomson et al. 2011) and RCP 8.5 (*red*, highest curve) (Riahi et al. 2011) as well as various other "blended" scenarios for CH$_4$ that are linear combinations of RCP 4.5 and RCP 8.5; (**b**) probability that the rise in ΔT in 2060 stays below 1.5 °C (*gold diamonds*) and 2.0 °C (*blue squares*) relative to pre-industrial, computed using RCP 4.5 combined with one of the blended CH$_4$ scenarios and plotted as a function of CH$_4$ in 2060; (**c**) same as (**b**), except for year 2100

sumption of more plant based foods (Stehfest et al. 2009; Pierrehumbert and Eshel 2015). The RCP 4.5 scenario (Thomson et al. 2011) projects a CH$_4$ mixing ratio of 1.80 ppm in 2060 (Table 4.2; Fig. 4.12), close to today's value. The RCP 4.5 design entails the use of a market pricing mechanism to stabilize global emissions of CH$_4$ (and CO$_2$, as well) at close to present level. Finally, the RCP 8.5 projection has atmospheric CH$_4$ at 2.7 ppm in 2060 and continuing to rise, unabated, until end of century (Fig. 4.12).

As noted in the introduction, the transition of production of electricity from coal to natural gas[32] is touted by many as being climate friendly because combustion of

---

[32] We use natural gas and methane interchangeably, but we are well aware that the mixture of gas burned in a so-called natural gas facility does contains trace amounts of hydrocarbons other than methane. Indeed, the association of ethane with methane in atmospheric samples can help distinguish whether a pulse of methane is due to ruminants or industry.

CH_4 yields 70 % more energy than combustion of coal, per molecule of CO_2 released to the atmosphere. The Clean Power Plan (CPP) proposed by the US Environmental Protection Agency[33] places limits on the abundance of CO_2 that can be emitted by electric generating units (EGUs) within each of the 50 states, by year 2030. To make a long story short, the US CPP facilitates the large-scale transition away from coal-fired EGUs to either natural gas or renewable EGUs. At time of writing, the US CPP is still being litigated. Of course, this policy is driven by the availability of a large domestic supply of CH_4 that is produced by horizontal fracturing of shale gas (i.e., fracking). Throughout the US, aging coal-fired EGUs are being replaced by new natural gas facilities, such as a 990 MW natural gas EGU scheduled to open in Brandywine, Maryland during 2018.[34] We mention the Brandywine plant to emphasize that in the US, market forces are driving replacement of coal-fired EGUs with natural gas units. Globally, however, atmospheric release of CO_2 from coal has been growing faster than atmospheric release of CO_2 from methane (Fig. 4.1).

The leakage of CH_4, at any point from extraction to just prior to combustion, tips the scales towards natural gas imposing a climate penalty rather than providing climate benefit (Howarth et al. 2011). Upon consideration of the latest values for the GWP of CH_4 from IPCC (2013), the break-even points are leakage of 6.9 % CH_4 for GWP on a 100-year time horizon and leakage of 2.3 % CH_4 for GWP on a 20-year time horizon.[35]

Choice of time horizon for the GWP of CH_4 is critical for deciding whether fracking in the US provides climate benefit or imposes climate penalty (Howarth 2014; Brandt et al. 2014). Table 4.4 shows estimates of the percentage of CH_4 leaked from active production sites, relative to daily production rates, from six selected recent studies that sample a large majority of the active natural gas extraction locations in the US. There is large variability in the estimated leakage rates. Regardless, one would conclude a more dire situation exists, with the climate balance likely swinging towards a penalty for fracking, upon use of the 2.3 % leakage rate tipping point (Howarth 2014). Conversely, one would conclude an overall climate benefit from fracking upon use of the 6.9 % break-even point and some of the measured

---

[33] https://www.epa.gov/cleanpowerplan/clean-power-plan-existing-power-plants#CPP-final

[34] This facility, described at http://www.pandafunds.com/invest/mattawoman, is not far from the where the authors of this book reside.

[35] The break-even point calculation is as follows. For each molecule of CO_2 released to the atmosphere, combustion of CH_4 yields 70 % more energy than combustion of coal. But, this benefit is potentially mitigated by release of an unknown amount of CH_4. For this calculation, we must use the GWP of CH_4 on a per molecule basis rather than a per mass basis (see footnote 8 of Chap. 1 for the distinction), because we are tracking release of molecules of CH_4 versus CO_2 to the atmosphere. Considering the per-molecule GWP of CH_4 on a 100 year time horizon of 10.2, we write:

$$CO_2 + \text{Unknown} \times 10.2 \times CO_2 = 1.7 \times CO_2$$

which yields Unknown = 0.069, or 6.9 % leakage for break-even. Use of the per-molecule GWP for CH_4 on a 20 year time horizon of 30.5 yields 2.3 % for the break even. Note to the experts: yes, we have not adjusted the right hand side of the equation for loss of energy that would have been put into the grid by the small amount of CH_4 that leaked. But this is more than offset by the presence in the natural gas system of a small amount of hydrocarbons that release more energy when burned, per CO_2 molecule released to the atmosphere, than is released by the combustion of CH_4.

Table 4.4 Estimates of % of CH$_4$ leakage relative to production in the US, selected studies

Leakage (%)	Region	Method	Citation
4.2–8.4	Bakken Shale, North Dakota	Aircraft sampling	Peischl et al. (2016)
1.0–2.1	Haynesville Shale, Louisiana and Texas	Aircraft sampling	Peischl et al. (2015)
1.0–2.8	Fayetteville Shale, Arkansas		
0.18–0.41	Marcellus Shale, Pennsylvania		
9.1 ± 6.2	Eagle Ford, Texas	Satellite sampling	Schneising et al. (2014)
10.1 ± 7.3	Bakken Shale, North Dakota		
0.42	190 production sites including Gulf Coast, Rocky Mountain, and Appalachia	In situ within facility grounds	Allen et al. (2013)
6.2–7.7	Unitah County, Utah	Aircraft sampling	Karion et al. (2013)
2.3–7.7	Julesburg Basin, Denver, Colorado	Tall tower and ground level mobile sampling	Pétron et al. (2012)

leakage rates given in Table 4.4 (Brandt et al. 2014). Quantification of leakage of CH$_4$ from production facilities will continue for quite some time, as will the debate regarding which leakage rate threshold should be used, as the community attempts to obtain consensus on whether fracking is friend or foe to climate. In some sense, we'd prefer to use GWP over a ~45-year time horizon, since our primary focus is projection of global warming out to 2060. We also direct the interested reader to a critique of the concept of GWP that suggests alternative metrics (Pierrehumbert 2014), which should be considered by those assessing CH$_4$ leakage from fracking.

Here we use another approach to assess the impact of CH$_4$ on the Paris Climate Agreement. The future projections of CH$_4$ offered by RCP 4.5 (Thomson et al. 2011) and RCP 8.5 (Riahi et al. 2011) are vastly different. Figure 4.12a compares these two projections along with various "blended" scenarios, which are linear combinations of the two extremes. Simulations of the future rise in ΔT have been conducted in the EM-GC framework for the six CH$_4$ scenarios shown in Fig. 4.12a; all other GHG and aerosol precursor values are based on RCP 4.5. Not only do these calculations provide a means for assessing the importance of controlling CH$_4$ leakage, but they also serve as a surrogate for quantifying the importance of future release of CH$_4$ from Arctic permafrost (Koven et al. 2011) (provided, of course, that atmospheric CH$_4$ stays bounded by the two extremes shown in Fig. 4.12a). As noted in Chap. 1, the present source of CH$_4$ from Arctic permafrost is small on a global scale (Kirschke et al. 2013), but this could change due to feedbacks in the climate system (Koven et al. 2011).

Figures 4.12b, c quantify the impact of future levels of atmospheric CH$_4$ on achieving the Paris thresholds. Figure 4.12b shows the cumulative probability that ΔT in year 2060, ΔT$_{2060}$, will remain below the Paris target of 1.5 °C (gold diamonds) or the Paris upper limit of 2.0 °C (blue squares). Figure 4.12c shows similar projections, but for 2100. Results are plotted as a function of the atmospheric mixing ratio of CH$_4$ for the respective end year. Otherwise, the calculations are calculated in an identical manner to that described for the EM-GC, RCP 4.5 entry in Table 2.1. The symbols associated with lowest value of CH_4 shown in Fig. 4.12b, c have the same numerical values as the appropriate entries for the EM-GC, RCP 4.5 row of Table 2.1.

The cumulative probabilities shown in Fig. 4.12 illustrate the importance of controlling future levels of atmospheric CH_4. If the goal is to achieve the Paris target of 1.5 °C warming, the EM-GC calculations suggest a ~79 % probability this will happen out to 2060, and a ~75 % probability out to 2100, if all GHGs follow RCP 4.5. If atmospheric CH_4 rises dramatically along the RCP 8.5 route and the future atmospheric abundance CO_2 falls along the RCP 4.5 trajectory, then the respective probabilities for achieving the Paris target decline to 58.3 % and 43.3 % in 2060 and 2100, respectively. Quantification of the impact of CH_4 on achieving the Paris target is provided for various other pathways, based on the points that lie in between the far left (RCP 4.5) and far right (RCP 8.5) entries. Finally, if atmospheric CO_2 can indeed be placed along the RCP 4.5 trajectory, then the EM-GC calculations indicate atmospheric CH_4 will likely not interfere with keeping global warming below the Paris 2.0 °C upper limit (Fig. 4.12c).

4.5 Paris Climate Agreement, Beacon of Hope

Even though society has obtained enormous benefit from the energy released by the combustion of fossil fuels, the relation between human activity, rising CO_2, and global warming is demonstrably clear (Chap. 1). We have used our Empirical Model of Global Climate to show that, if future abundances of CO_2, CH_4, and N_2O follow the trajectory of the RCP 4.5 scenario (Thomson et al. 2011), there is greater than 95 % probability the rise in global mean surface temperature during the rest of this century will stay below 2 °C warming (relative to pre-industrial baseline) and a ~75 % chance future warming will stay below 1.5 °C warming (Chap. 2). Our analysis of the INDCs that constitute the Paris Climate Agreement (Chap. 3) show that GHG emissions will remain below RCP 4.5 out to 2060 if:

(1) conditional as well as unconditional pledges are met
(2) reductions in GHG emissions needed to achieve the Paris commitments, which generally extend to 2030, are propagated forward to 2060

The Paris Climate Agreement, as presently constituted, provides a beacon of hope that climate catastrophe can be avoided.

The Paris INDCs, with rare exception, extend only to 2030. It is essential the world begin planning for a 2060 future. Market forces, driven by the low cost availability of natural gas, will facilitate achievement or near achievement of the Paris commitments in some nations, such as the US, without particularly a aggressive transition to renewable energy. However, the gap between market driven production of energy by the combustion of fossil fuels and the limit of RCP 4.5 grows dramatically between 2030 and 2060 (Fig. 4.2). Assuming a 5.7 % share of nuclear energy in 2060, then achievement of RCP 4.5 emissions of CO_2 requires 50 % of global energy to be produced by renewables in 2060 (Fig. 4.3). The global climate models used by IPCC (IPCC 2013) indicate the RCP 2.6 emission trajectory has to be followed to keep warming below 2.0 °C. If this is indeed true, then 88 % of the global demand of energy by 2060 will need to be produced by methods with negligible impact on atmospheric GHGs (Fig. 4.5).

Many communities, towns, and nations have embraced the challenge. Green Mountain College in the state of Vermont has a credible plan in place to obtain 100 % of the energy consumed on campus by renewable sources in 2020.[36] The central element of this effort is a biomass plant that uses locally harvested woodchips to generate heat. Samsö, Denmark, an island of about 4000 inhabitants, has a net negative carbon footprint thanks to 22 massive wind turbines, most of which are owned by members of the community.[37] Germany is planning to increase the share of total, nation-wide energy consumed that is provided by renewables from 12.6 % in 2015 to 60 % by 2050.[38] The German effort is multifaceted, involving various forms of renewable energy as well as state-of-the-art building efficiency standards. In Sect. 4.3, solar energy projects in Nigeria, Rwanda, and South Africa were described. It is incumbent the rest of the world embrace and emulate the efforts of Green Mountain College, Samsö, Germany, the solar projects in Africa, and so many other communities, towns, and nations that are actively transitioning to renewables. Fifty percent of *total global energy* by renewables in year 2060 is a very tall pole. As populations expand and standards of living rise, 50 % renewables in 2060 will be needed to have a reasonably good chance of achieving the goals of the Paris Climate Agreement.

4.6 Methods

Many of the figures use data or archives of model output from publically available sources. Here, webpage addresses of these archives, citations, and details regarding how data and model output have been processed are provided. Only those figures with "see methods for further information" in the caption are addressed below. Electronic copies of the figures are available on-line at http://parisbeaconofhope.org.

Figure 4.1 shows time series for emissions of atmospheric CO_2 from land use change, combustion of solid (coal), liquid (petroleum), and gaseous (methane) forms of fossil fuel, as well as cement production and gas flaring. The data for emissions of CO_2 from the combustion of fossil fuels (Boden et al. 2013) and land use change (Houghton et al. 2012) originate from two files hosted by the Carbon

---

[36] http://www.greenmtn.edu/sustainability/sustainability-2020

[37] http://www.scientificamerican.com/article/samso-attempts-100-percent-renewable-power

[38] https://www.cleanenergywire.org/factsheets/germanys-greenhouse-gas-emissions-and-climate-targets

Dioxide Information Analysis Center (CDIAC) at the US Department of Energy's (DOE) Oak Ridge National Laboratory (ORNL):

http://cdiac.ornl.gov/ftp/ndp030/global.1751_2013.ems

http://cdiac.ornl.gov/trends/landuse/houghton/1850-2005.txt

http://cdiac.ornl.gov/ftp/Global_Carbon_Project/Global_Carbon_Budget_2015_v1.1.xlsx

The first file was used for CO_2 emissions from 1850 to 1958 for all sources other than land use change; the second file was used for CO_2 emissions from land use change from 1850 to 1958; and the third file was used for all of the emissions from 1959 to 2014.

Figure 4.2 shows global energy consumption and CO_2 emissions from the US EIA (1990–2040) and a linear extrapolation of these values out to 2060. Data in Fig. 4.2a are from:

http://www.eia.gov/forecasts/ieo/excel/figurees2_data.xls

and data in Fig. 4.2b originate from:

http://www.eia.gov/forecasts/ieo/excel/figurees8_data.xls

We have extrapolated the EIA values to 2060 by conducting a linear fit to each component on both panels, using data from 2030 to 2040, and propagating forward to 2060 using the slope and intercept of each fit. Figure 4.2b also contains estimates of GHG emissions from RCP 4.5 and 8.5 (blue and red lines). These estimates are based on files hosted by the Potsdam Institute for Climate Research (PICR) (Meinshausen et al. 2011) at:

http://www.pik-potsdam.de/~mmalte/rcps/data/RCP45_EMISSIONS.DAT

http://www.pik-potsdam.de/~mmalte/rcps/data/RCP85_EMISSIONS.DAT

The emissions of CO_2 in the RCP files include sources from combustion of fossil fuels, cement, and gas flaring. The EIA emissions shown in Fig. 4.2b are only for combustion of fossil fuels. We have adjusted the RCP emissions to ensure an apples to apples comparison with the EIA-based estimate by: (1) computing the ratio from CDIAC data of [fossil fuel emissions of CO_2]/[fossil fuel + cement + flaring emissions of CO_2] for years 1990–2014; (2) extrapolating this ratio to 2060 using a linear fit, since it exhibits a modest, steady linear decline over time; (3) multiplying the RCP emissions by our linear fit to the extrapolated ratio. The ratio used to multiply the RCP emissions equals 0.94 in 2013 and 0.88 in 2060: i.e., the adjustment is modest in all years.

Figure 4.3 shows global energy consumption and emissions of CO_2 by source, modified to meet RCP 4.5 emissions of CO_2 starting in year 2030 (Thomson et al. 2011). The sources of data and adjustment to the RCP emissions (blue and red lines), to account for cement product and gas flaring, are handled in the same manner as described above for Fig. 4.2. We have forced the sum of CO_2 emitted by coal, natural gas, and liquid fuels shown by the three colored wedges in Fig. 4.3b to match the global emission of CO_2 from RCP 4.5 starting in year 2030 by preserving the percentage contribution of energy supply from coal, natural gas, and liquid fuels relative to the sum of these three quantities, for each year. Since the release of CO_2 from these three sources is projected to exceed RCP 4.5 in year 2020, we linearly interpolate the value in 2020, to 2030, to provide a smooth transition to the 2030 match. In order to obtain the new energy consumption estimates from coal, natural gas, and liquid fuels shown in Fig. 4.3a, we have used the ratio of quad BTU per year provided from each source divided by the Gt CO_2 released from each source, extracted from the EIA baseline projection. For those keeping score at home, in year 2040 these three ratios are 18.90, 15.82, and 10.93 quad BTU/Gt CO_2 for natural gas, liquid fuels, and coal, respectively. Finally, the shortfall in meeting the global demand for energy, caused by the decline in fossil fuel-based production needed to meet RCP 4.5 emission of CO_2, is assigned to the renewables category.

Figure 4.4 shows energy production by renewables broken into two wedges: energy from the combustion of biomass (dark green) and other sources of renewable energy (light green). The sum of the biomass and other wedges in Fig. 4.4a matches, by design, the renewable time series shown in Fig. 4.2a. The numerical values of energy production from renewables other than biomass shown in Fig. 4.4a are based on electricity production from renewables projections for 2012 to 2040 in:

http://www.eia.gov/forecasts/ieo/excel/figure5-4_data.xls

For years 2001–2011, the apportionment of biomass versus others was based on information archived in Organization for Economic Co-operation and Development (OECD), International Energy Administration (IEA) annual Renewables Information reports, available on the web at:

http://www.oecd-ilibrary.org/energy/renewables-information_20799543.

For 2000 and years prior, we assumed 80 % of total global energy from renewables had been supplied by biomass, which is the IEA percentage for 2001 and 2002. For Fig. 4.4b, we assign to "other" all of the new energy from renewables needed to match RCP 4.5 emissions of CO_2. The size of this wedge is enormous: it exceeds the world demand for electricity in the latter years. For this reason, this wedge is labeled "renewable sources of energy for industry, electricity, transportation, and building".

Figure 4.5 is similar to Fig. 4.3, except the target for emissions of CO_2 is RCP 2.6 (van Vuuren et al. 2011). The RCP2.6 emissions originate from file:

http://www.pik-potsdam.de/~mmalte/rcps/data/RCP3PD_EMISSIONS.DAT

The use of 3PD in this filename is due to the fact some researchers had called this scenario RCP 3 Peak and Decline, rather than RCP 2.6. The analysis procedure (i.e., adjustment for CO_2 from cement and gas flaring; preservation of the EIA ratios of coal, natural gas, and liquid fuels, etc.) is the same as described above. For illustrative purposes, and since the design of RCP 2.6 (van Vuuren et al. 2011) mentions carbon capture and sequestration (CCS) whereas the design of RCP 4.5 (Thomson et al. 2011) does not consider this still developing technology, we have kept the renewable wedge in Fig. 4.5a the same as the renewable wedge in Fig. 4.3a, and assigned to CCS the new shortfall needed to achieve RCP 2.6 emissions of CO_2.

Figure 4.6 shows global maps of population and night lights. The population data (Doxsey-Whitfield et al. 2015) were obtained from file:

gpw-v4-population-density-adjusted-to-2015-unwpp-country-totals-2015.zip

downloaded from the NASA SEDAC website at:

http://beta.sedac.ciesin.columbia.edu/data/set/gpw-v4-population-density-adjusted-to-2015-unwpp-country-totals/data-download

This dataset is provided at 30-arcsecond resolution, in units of people per square km (ppl km^{-2}). The appropriate 225 data points (15 × 15) were averaged to produce population on the 0.125° × 0.125° (latitude, longitude) grid used for the figure. The night lights data consist of measurements in a series of files downloaded from the VIIRS DNB Cloud Free Composites tab at:

http://ngdc.noaa.gov/eog/viirs.html

The VCMSLCFG series of data files, which do not include stray light correct, were used because the stray light corrected product has greatly reduced data coverage at high latitudes. This raw product is provided on a 15-arcsecond grid. The appropriate 900 data points (30 × 30) were averaged to produce night lights data on our 0.125° × 0.125° (latitude, longitude) grid.

The raw VIIRS product contains considerable signals from aurora borealis and fires. Since it is our objective to show night lights data representative of electricity, we have filtered the data to remove aurora and fires. The obvious aurora signals occurred poleward of 42°N in the NH, poleward of 40°S in the western part of the SH, and poleward of 50°S in the eastern part of the SH. For these regions, we set the night lights value to zero if the corresponding population was below 5 ppl km^{-2}.

The contribution to VIIRS night lights from fires was removed using the NASA MODIS monthly fire count product (Giglio et al. 2006) for 2015, downloaded from:

ftp://neespi.gsfc.nasa.gov/data/s4pa/Fire/MOD14CM1.005/2015

Monthly fire count data are available at 1° × 1° (latitude, longitude). Monthly fire count data are averaged to produce an annual field at 1° × 1°. If a cell has an annual average fire count value larger than 5, this indicates the VIIRS signal was likely influenced by an active fire. In this case, since fires are seasonal, the night lights values for each 0.125° × 0.125° cell within the fire affected 1° × 1 grid was replaced with the minimum night lights value observed by VIIRS over the course of 2015. These simple methods to remove the influence of aurora and fire led to an obvious, dramatic improvement in the rendering of light likely due to the availability of electricity, based on visual inspection of before and after images together with population density maps.

Figure 4.9 shows scatter plots of night lights versus population. Data are only shown if population of the 0.125° × 0.125° grid exceeds 5 ppl km^{-2}. The observations for each region were sorted, from lowest to highest population. The sorted data was then divided into twenty bins, all with the same (or nearly the same) number of data. Once sorted and binned, we then computed the median population for each set, as well as the 5th, 25th, 50th (median), 75th, and 95th percentile of the night lights distribution. The figure shows the raw data (speckles) and each percentile, as described in the caption. For Africa, most of the night lights measurements fell below the lower end of the vertical axis; only the 95th percentile, 75th percentile, and a single median point for Africa lies within the range of the vertical axis.

Figure 4.10 shows estimates of TCRE from CMIP5 GCMs and our EM-GC. The EM-GC simulations are based on a single run for each RCP scenario. We have written extensively about all of the RCP scenarios besides RCP 6.0. Mixing ratios of CO_2, CH_4, and N_2O for RCP 6.0 (Masui et al. 2011) are shown in Fig. 2.1, and files used to drive the EM-GC calculation for this scenario were obtained from PICR (Meinshausen et al. 2011) at:

http://www.pik-potsdam.de/~mmalte/rcps/data

Figure 4.11 shows probabilistic estimates of TCRE found using our EM-GC, constrained by RCP 8.5 emissions. All simulations use OHC based on the average of six data records shown in Fig. 2.8, and have been weighted by $1/\chi^2$ prior to calculation of the probabilities. Cumulative CO_2 emissions due to land use change, combustion of fossil fuel, cement production, and flaring from RCP 8.5 (Riahi et al. 2011), as archived by PICR at the link given for Methods of Fig. 4.2, are used to define the horizontal axis.

References

Allen MR, Frame DJ, Huntingford C, Jones CD, Lowe JA, Meinshausen M, Meinshausen N (2009) Warming caused by cumulative carbon emissions towards the trillionth tonne. Nature 458(7242):1163–1166,   http://www.nature.com/nature/journal/v458/n7242/suppinfo/nature08019_S1.html

Allen DT, Torres VM, Thomas J, Sullivan DW, Harrison M, Hendler A, Herndon SC, Kolb CE, Fraser MP, Hill AD, Lamb BK, Miskimins J, Sawyer RF, Seinfeld JH (2013) Measurements of methane emissions at natural gas production sites in the United States. Proc Natl Acad Sci 110(44):17768–17773. doi:10.1073/pnas.1304880110

Bauer N, Mouratiadou I, Luderer G, Baumstark L, Brecha RJ, Edenhofer O, Kriegler E (2016) Global fossil energy markets and climate change mitigation—an analysis with REMIND. Clim Chang 136(1):69–82. doi:10.1007/s10584-013-0901-6

Boden TA, Marland G, Andres RJ (2013) Global, regional, and national fossil-fuel CO_2 emissions. Oak Ridge National Laboratory, Oak Ridge, TN. doi:10.3334/CDIAC/00001_V2013

Bonan GB (1999) Frost followed the plow: impacts of deforestation on the climate of the United States. Ecol Appl 9(4):1305–1315. doi:10.1890/1051-0761(1999)009[1305:FFTPIO]2.0.CO;2

Brandt AR, Heath GA, Kort EA, O'Sullivan F, Pétron G, Jordaan SM, Tans P, Wilcox J, Gopstein AM, Arent D, Wofsy S, Brown NJ, Bradley R, Stucky GD, Eardley D, Harriss R (2014) Methane leaks from North American natural gas systems. Science 343(6172):733–735. doi:10.1126/science.1247045

Canty T, Mascioli NR, Smarte MD, Salawitch RJ (2013) An empirical model of global climate—Part 1: A critical evaluation of volcanic cooling. Atmos Chem Phys 13(8):3997–4031. doi:10.5194/acp-13-3997-2013

Crutzen PJ, Mosier AR, Smith KA, Winiwarter W (2008) N_2O release from agro-biofuel production negates global warming reduction by replacing fossil fuels. Atmos Chem Phys 8(2):389–395

Dlugokencky EJ, Bruhwiler L, White JWC, Emmons LK, Novelli PC, Montzka SA, Masarie KA, Lang PM, Crotwell AM, Miller JB, Gatti LV (2009) Observational constraints on recent increases in the atmospheric CH_4 burden. Geophys Res Lett 36(18):L18803. doi:10.1029/2009GL039780

Doxsey-Whitfield E, MacManus K, Adamo SB, Pistolesi L, Squires J, Borkovska O, Baptista SR (2015) Taking advantage of the improved availability of census data: a first look at the gridded population of the world, version 4. Papers Appl Geogr 1(3):226–234. doi:10.1080/23754931.2015.1014272

Fearnside PM (2002) Greenhouse gas emissions from a hydroelectric reservoir (Brazil's Tucuruí Dam) and the energy policy implications. Water Air Soil Pollut 133(1):69–96. doi:10.1023/a:1012971715668

Giglio L, Csiszar I, Justice CO (2006) Global distribution and seasonality of active fires as observed with the Terra and Aqua Moderate Resolution Imaging Spectroradiometer (MODIS) sensors. J Geophys Res Biogeosci 111(G2):G02016. doi:10.1029/2005JG000142

Gunkel G (2009) Hydropower—a green energy? Tropical reservoirs and greenhouse gas emissions. CLEAN Soil Air Water 37(9):726–734. doi:10.1002/clen.200900062

Hammond GP, Spargo J (2014) The prospects for coal-fired power plants with carbon capture and storage: a UK perspective. Energy Convers Manage 86:476–489, http://dx.doi.org/10.1016/j.enconman.2014.05.030

Hillger D, Kopp T, Lee T, Lindsey D, Seaman C, Miller S, Solbrig J, Kidder S, Bachmeier S, Jasmin T, Rink T (2013) First-light imagery from Suomi NPP VIIRS. Bull Am Meteorol Soc 94(7):1019–1029. doi:10.1175/BAMS-D-12-00097.1

Houghton RA, House JI, Pongratz J, van der Werf GR, DeFries RS, Hansen MC, Le Quéré C, Ramankutty N (2012) Carbon emissions from land use and land-cover change. Biogeosciences 9(12):5125–5142. doi:10.5194/bg-9-5125-2012

Howarth RW (2014) A bridge to nowhere: methane emissions and the greenhouse gas footprint of natural gas. Energy Sci Eng 2(2):47–60. doi:10.1002/ese3.35

Howarth RW, Santoro R, Ingraffea A (2011) Methane and the greenhouse-gas footprint of natural gas from shale formations. Clim Chang 106(4):679–690. doi:10.1007/s10584-011-0061-5

IPCC (2005) IPCC special report on carbon dioxide capture and storage. Cambridge University Press for the Intergovernmental Panel on Climate Change, Cambridge

IPCC (2013) Climate change 2013: the physical science basis. contribution of working group I to the fifth assessment report of the intergovernmental panel on climate change. Cambridge, UK and New York, NY, USA

Jakob M, Hilaire J (2015) Climate science: unburnable fossil-fuel reserves. Nature 517(7533):150–152. doi:10.1038/517150a

Jones PD, Lister DH, Osborn TJ, Harpham C, Salmon M, Morice CP (2012) Hemispheric and large-scale land-surface air temperature variations: an extensive revision and an update to 2010. J Geophys Res 117(D5):D05127. doi:10.1029/2011jd017139

Karion A, Sweeney C, Pétron G, Frost G, Michael Hardesty R, Kofler J, Miller BR, Newberger T, Wolter S, Banta R, Brewer A, Dlugokencky E, Lang P, Montzka SA, Schnell R, Tans P, Trainer M, Zamora R, Conley S (2013) Methane emissions estimate from airborne measurements over a western United States natural gas field. Geophys Res Lett 40(16):4393–4397. doi:10.1002/grl.50811

Kirschke S, Bousquet P, Ciais P, Saunois M, Canadell JG, Dlugokencky EJ, Bergamaschi P, Bergmann D, Blake DR, Bruhwiler L, Cameron-Smith P, Castaldi S, Chevallier F, Feng L, Fraser A, Heimann M, Hodson EL, Houweling S, Josse B, Fraser PJ, Krummel PB, Lamarque J-F, Langenfelds RL, Le Quere C, Naik V, O'Doherty S, Palmer PI, Pison I, Plummer D, Poulter B, Prinn RG, Rigby M, Ringeval B, Santini M, Schmidt M, Shindell DT, Simpson IJ, Spahni R, Steele LP, Strode SA, Sudo K, Szopa S, van der Werf GR, Voulgarakis A, van Weele M, Weiss RF, Williams JE, Zeng G (2013) Three decades of global methane sources and sinks. Nat Geosci 6(10):813–823. doi:10.1038/ngeo1955, http://www.nature.com/ngeo/journal/v6/n10/abs/ngeo1955.html#supplementary-information

Kopetz H (2013) Renewable resources: build a biomass energy market. Nature 494(7435):29–31

Koven CD, Ringeval B, Friedlingstein P, Ciais P, Cadule P, Khvorostyanov D, Krinner G, Tarnocai C (2011) Permafrost carbon-climate feedbacks accelerate global warming. Proc Natl Acad Sci 108(36):14769–14774. doi:10.1073/pnas.1103910108

Liao LB, Weiss S, Mills S, Hauss B (2013) Suomi NPP VIIRS day-night band on-orbit performance. J Geophys Res 118(22):12,705–712,718. doi:10.1002/2013JD020475

Lynas M (2008) Six degrees: our future on a hotter planet. National Geographic, Washington, DC

MacDougall AH, Friedlingstein P (2015) The origin and limits of the near proportionality between climate warming and cumulative CO_2 emissions. J Clim 28(10):4217–4230. doi:10.1175/JCLI-D-14-00036.1

Masui T, Matsumoto K, Hijioka Y, Kinoshita T, Nozawa T, Ishiwatari S, Kato E, Shukla PR, Yamagata Y, Kainuma M (2011) An emission pathway for stabilization at 6 W m^{-2} radiative forcing. Clim Chang 109(1–2):59–76. doi:10.1007/s10584-011-0150-5

Meinshausen M, Smith SJ, Calvin K, Daniel JS, Kainuma MLT, Lamarque JF, Matsumoto K,
    Montzka SA, Raper SCB, Riahi K, Thomson A, Velders GJM, Vuuren DPP (2011) The RCP
    greenhouse gas concentrations and their extensions from 1765 to 2300. Clim Chang 109(1–2):
    213–241. doi: 10.1007/s10584-011-0156-z
Muench S, Guenther E (2013) A systematic review of bioenergy life cycle assessments. Appl
    Energy 112:257–273, http://dx.doi.org/10.1016/j.apenergy.2013.06.001
NRC (2015) Climate intervention: carbon dioxide removal and reliable sequestration. National
    Academies Press, Washington, DC
Olah GA, Goeppert A, Prakash GKS (2009) Beyond oil and gas: the methanol economy. 2nd edn.
    Wiley-VCH, Weinheim [an der Bergstrasse, Germany]
Peischl J, Ryerson TB, Aikin KC, de Gouw JA, Gilman JB, Holloway JS, Lerner BM, Nadkarni R,
    Neuman JA, Nowak JB, Trainer M, Warneke C, Parrish DD (2015) Quantifying atmospheric
    methane emissions from the Haynesville, Fayetteville, and northeastern Marcellus shale gas
    production regions. J Geophys Res Atmos 120(5):2119–2139. doi:10.1002/2014JD022697
Peischl J, Karion A, Sweeney C, Kort EA, Smith ML, Brandt AR, Yeskoo T, Aikin KC, Conley
    SA, Gvakharia A, Trainer M, Wolter S, Ryerson TB (2016) Quantifying atmospheric methane
    emissions from oil and natural gas production in the Bakken shale region of North Dakota.
    J Geophys Res Atmos 121(10):6101–6111. doi:10.1002/2015JD024631
Pétron G, Frost G, Miller BR, Hirsch AI, Montzka SA, Karion A, Trainer M, Sweeney C, Andrews
    AE, Miller L, Kofler J, Bar-Ilan A, Dlugokencky EJ, Patrick L, Moore CT, Ryerson TB, Siso
    C, Kolodzey W, Lang PM, Conway T, Novelli P, Masarie K, Hall B, Guenther D, Kitzis D,
    Miller J, Welsh D, Wolfe D, Neff W, Tans P (2012) Hydrocarbon emissions characterization in
    the Colorado Front Range: a pilot study. J Geophys Res Atmos 117(D4):D04304. doi:10.1029
    /2011JD016360
Pierrehumbert RT (2014) Short-lived climate pollution. Annu Rev Earth Planet Sci 42(1):341–379.
    doi:10.1146/annurev-earth-060313-054843
Pierrehumbert RT, Eshel G (2015) Climate impact of beef: an analysis considering multiple time
    scales and production methods without use of global warming potentials. Environ Res Lett
    10(8):085002
Pimentel D (2012) Global economic and environmental aspects of biofuels. CRC Press, Boca
    Raton
Rathmann R, Szklo A, Schaeffer R (2010) Land use competition for production of food and liquid
    biofuels: an analysis of the arguments in the current debate. Renew Energy 35(1):14–22, http://
    dx.doi.org/10.1016/j.renene.2009.02.025
Riahi K, Rao S, Krey V, Cho C, Chirkov V, Fischer G, Kindermann G, Nakicenovic N, Rafaj P
    (2011) RCP 8.5—a scenario of comparatively high greenhouse gas emissions. Clim Chang
    109(1–2):33–57. doi:10.1007/s10584-011-0149-y
Rogelj J, den Elzen M, Höhne N, Fransen T, Fekete H, Winkler H, Schaeffer R, Sha F, Riahi K,
    Meinshausen M (2016a) Paris Agreement climate proposals need a boost to keep warming well
    below 2 °C. Nature 534(7609):631–639. doi:10.1038/nature18307, http://www.nature.com/
    nature/journal/v534/n7609/abs/nature18307.html#supplementary-information
Rogelj J, Schaeffer M, Friedlingstein P, Gillett NP, van Vuuren DP, Riahi K, Allen M, Knutti R
    (2016b) Differences between carbon budget estimates unravelled. Nat Clim Change 6(3):
    245–252. doi:10.1038/nclimate2868
Schilmann A, Riojas-Rodríguez H, Ramírez-Sedeño K, Berrueta VM, Pérez-Padilla R, Romieu I
    (2015) Children's respiratory health after an efficient biomass stove (Patsari) intervention.
    EcoHealth 12(1):68–76. doi:10.1007/s10393-014-0965-4
Schneising O, Burrows JP, Dickerson RR, Buchwitz M, Reuter M, Bovensmann H (2014) Remote
    sensing of fugitive methane emissions from oil and gas production in North American tight
    geologic formations. Earth's Future 2(10):548–558. doi:10.1002/2014EF000265
Shi K, Yu B, Huang Y, Hu Y, Yin B, Chen Z, Chen L, Wu J (2014) Evaluating the ability of NPP-
    VIIRS nighttime light data to estimate the gross domestic product and the electric power con-

sumption of China at multiple scales: a comparison with DMSP-OLS data. Remote Sens 6(2):1705

Spigarelli BP, Kawatra SK (2013) Opportunities and challenges in carbon dioxide capture. J CO_2 Utili 1:69–87. http://dx.doi.org/10.1016/j.jcou.2013.03.002

Stehfest E, Bouwman L, van Vuuren DP, den Elzen MGJ, Eickhout B, Kabat P (2009) Climate benefits of changing diet. Clim Chang 95(1):83–102. doi:10.1007/s10584-008-9534-6

Taylor KE, Stouffer RJ, Meehl GA (2012) An overview of CMIP5 and the experiment design. Bull Am Meteorol Soc 93(4):485–498. doi:10.1175/bams-d-11-00094.1

Thomson AM, Calvin KV, Smith SJ, Kyle GP, Volke A, Patel P, Delgado-Arias S, Bond-Lamberty B, Wise MA, Clarke LE, Edmonds JA (2011) RCP4.5: a pathway for stabilization of radiative forcing by 2100. Clim Chang 109(1–2):77–94. doi:10.1007/s10584-011-0151-4

van Vuuren DP, Stehfest E, Elzen MGJ, Kram T, Vliet J, Deetman S, Isaac M, Klein Goldewijk K, Hof A, Mendoza Beltran A, Oostenrijk R, Ruijven B (2011) RCP2.6: exploring the possibility to keep global mean temperature increase below 2 °C. Clim Chang 109(1–2):95–116. doi:10.1007/s10584-011-0152-3

Wickramasinghe A (2003) Gender and health issues in the biomass energy cycle: impediments to sustainable development. Energy Sustain Dev 7(3):51–61. http://dx.doi.org/10.1016/S0973-0826(08)60365-8

Index

Printed in the United States
By Bookmasters